PATER MUNDI;

OR,

MODERN SCIENCE TESTIFYING TO THE HEAVENLY FATHER.

BEING

IN SUBSTANCE

LECTURES DELIVERED TO SENIOR CLASSES IN AMHERST COLLEGE.

BY

REV. E. F. BURR, D. D.,

AUTHOR OF "ECCE CŒLUM."

Φήμη δ' οὔτις πάμπαν ἀπόλλυται, ἥντινα πολλοὶ
Λαοὶ φημίζουσι· Θεός νύ τίς ἐστι καὶ αὐτή.

Hesiod.

FIRST SERIES.

FIFTH EDITION.

BOSTON:
NOYES, HOLMES, AND COMPANY.
No. 117 Washington Street.
1872.

RIVERSIDE, CAMBRIDGE:
STEREOTYPED AND PRINTED BY
H. O. HOUGHTON AND COMPANY

To the

HEAVENLY FATHER,

TO WHOM WE DEDICATE OUR SABBATHS, OUR SANCTUARIES, AND OURSELVES,

These Volumes,

IN ILLUSTRATION OF HIS BEING AND GREATNESS,

ARE REVERENTLY INSCRIBED.

PREFACE.

THE whole plan of the author looks beyond the present volumes. It proposes to defend and illustrate both Theism and Christianity from the side of Modern Science. This accounts for the structure of the first two lectures.

In the second volume the appeals to the Sciences will be found more direct and full than even in this — especially as negativing that Law Scheme which is the only present competitor of Theism as an explanation of Nature.

These lectures were designed to be spoken to College Classes on the eve of graduation. Hence some peculiarities. They speak to the ear. They speak to the young. They speak to educated young men who may be presumed familiar with general classical as well as scientific knowledge; and whom it is of the last importance to have go forth into the world richly assured of the exceeding breadth of the Christian Foundations, and richly prepared to manifest them to all unbelievers. So the lectures are zealous for a side. They are anxious to carry a point. They appear not to have discovered that one must be indifferent in order to be fair. They affect no philosophic impartiality; but speak as a Christian believer, to the sons of Christian parents, and within a Christian college which has not yet thought it necessary to teach neutrality (or worse) between Christianity and Buddhism, from chairs resting on Christian endowments.

The author states some things very strongly. But he does not suppose himself to have stated them more strongly than facts warrant. He feels very hostile to Atheism. He holds it the worst enemy of mankind. Its recent attempts to shelter itself under the great name of Science greatly move his indignation. He is amazed at its effrontery in claiming that a single true science looks on it with favor. At the same time he aims to be just, even to Satan. What he would gladly destroy in the interest of humanity, he would only destroy by the lawful use of lawful weapons.

The larger part of the sixth lecture has been published before. But as it properly belongs to this course of lectures, and as the omission of it would, in the author's view, mar the symmetry of his general plan, he has thought best to insert it in its proper place.

LYME, CONN., Nov. 30, 1869.

CONTENTS.

V. In Tenebris.

VI. Harmonies with Nature.

VII. Need of God.

VIII. Theism as a Scientific Hypothesis.

I.

EXPERIMENTAL METHOD.

Ψηλαφήσατέ με χαι ἴδετε.

Καλῶς ἂν σοι ὁ Θεὸς Αὐτὸς ξυλλαμβάνοι. — *Plato.*

I. Experimental Method.

FIRST LECTURE.

EXPERIMENTAL METHOD.

THERE are two ways in which men assure themselves of the qualities of material objects. One is the way of argument: the other is that of direct personal experiment. A man of reputation tells me that a certain sort of wood is tough, flexible, and hard; or I see it extensively used for purposes to which these qualities are essential; or the general appearance and arrangement of the fiber, I find, are the same as in other woods known to have these qualities — these are so many arguments from which in a way of inference my mind reaches a belief in the toughness, flexibility, and hardness of that wood. But, if I choose, I may reach the same belief in another way. I may strike my own hammer on that wood, and see what resistance it makes to indentation. I may take it into my own hands and try to bend it. I may with my own fingers or wedges attempt to tear it asunder. Thus by a direct personal trial, and not at all in the way of argumentative inference, I may convince myself that the wood is what it is claimed to be.

In the same twofold way we may satisfy ourselves of the existence of certain spiritual qualities. Is your acquaintance generous, is he honest, is he capable? You may argue out an answer for yourself, or you may obtain it by the personal application of certain practical tests. Honest? Yes, you may say, for it is an honest family to which he belongs, and I know that from childhood he has had iustruction and training fitted to make an honest man. Besides, he bears a good reputation for honesty. Those with whom he has had dealings speak well of him. This is argument. A judgment is reached inferentially from other judgments or facts. But there is such a thing as your making a direct experiment on the man which will settle the question of his honesty to your mind without help from any other quarter. Put in his way an opportunity of taking some small unfair advantage of you with apparently entire safety, and see what he will do with it. Try him again and again at a variety of points, and watch how he carries himself under the temptation. This will finally show you what the man is — perhaps will show you that his word is as good as a bond, and that you might venture to trust him with every dollar you are worth. You have personally experimented upon him in true Baconian and scientific way, and found him trustworthy in the last degree. With your own hands you have applied the acid to what men call gold, and have found that gold it really is. See how the metal shines under the nitric drop!

Suppose, now, that our inquiries, instead of relating to attributes of matter or attributes of the human soul, relate to that still higher plane of thought, the attributes of God and His Word — say the reality of the Christian God and the divinity of the Christian Scriptures. Have we still the same two ways of information that are universally allowed in dealing with those questions of the lower order? Can we properly argue, and can we properly experiment also? The first question I reserve to be answered in the next lecture: the second I propose to answer now, because I regard it as primary in its character. I repeat, can we and may we put things of such great names and august claims as the Christian God and the Christian Scriptures under substantially just such direct practical tests as show us that a given wood is hard, and a given man honest? This question is an interesting one — for the reasons that the radical experimental method is found so enormously powerful and fruitful in the lower fields of inquiry, that we need all the light on the alleged God and Revelation we can possibly obtain, and that there is more or less current the idea that it is not possible, or at least lawful, to deal with such great spiritual matters in the way of critical experiment. The great questions that stand before the world from age to age, and which make all others almost invisible, are these. Is God real? Are the Christian Scriptures His message? There are some in the world — we suppose an ever-decreasing number — who to these questions are prepared to say, "No," or are not prepared to say, "Yes" — disbelievers or unbelievers. Then there is another class who truly believe in God and Scripture; but their faith is far from being as large-limbed, and muscular, and majestic of mien as they could desire. Lastly, there are those who themselves believe almost as though they saw, but who would like to communicate something of their own full assurance of faith to the many around whose condition is less happy, and on whom mere argument seems so largely spent in vain. To all these classes it is a question of very great moment whether the field of religion, like every other field, is open to the double-handed exploration of argument and personal experiment — whether, after having exhausted or, what is much better, before touching the system of premises and inferences, they may not bare their arms and go forth on the subjects of God and Scripture with such practical tests as shall be to them what the hammer is to the wood that asks to be considered hard, and actual opportunities of safe cheating to the man who asks to be considered honest. The idea of experimenting on God and His Word may have at first quite an objectionable look. It looks, perhaps, like irreverence and audacity and desecration. One gets his mind filled with the idea of coarse mechanical experiments and of the harsh, irreverent ways in which they are sometimes played off on creature-natures; and when mention is made of religious experiments, the gross old ideas still cling about the new thought. It seems as if nothing of the kind would be allowable out of the low realm of the commonplace and profane world. How it sounds to talk of trying experiments on God and Religion!

In answering this current, or at least not unfrequent, feeling, it must be admitted at the outset that there are experiments on these objects of which we may not entertain thought for a moment. They would be extreme presumption and sacrilege. Our instinctive sense of propriety would revolt from them as putting dishonor on the conception of a God and a Religion. When the Jews came to Jesus on a certain occasion, saying, "Master, we would see a sign from thee," what they proposed to do was then and there to try a direct experiment on His miraculous power. The proposal met a severe rebuff. If one of you should rise in his place and say, "If there is a God, let Him immediately show Himself by casting yonder hill into the river," his experiment would be a very wrong one. If one of you should take it on himself to cry out towards the heavens, "If the religion of Jesus is divine, let rain this moment fall from a clear sky," his experiment would be a very wrong one. If he should put Liberalism to a similar test, saying, "If it is really Scripture that God is Trinity and future punishment everlasting, let a plumed angel at once appear in that door-way, and say so," his experiment would be a very wrong one. All such tests are plain irreverence and presumption. They set up our wisdom as supreme, and presume to dictate terms and methods to God. This will never do. Let the rash man take the shoes from his feet as he nears the place where perchance God is concealed: why must a voice smite him with the information that all such places are holy?

Yes, there are many experiments on God and the Scriptures which would be highly improper — say, if you please, intolerable. But it would be a great misfortune if, on glancing at some of these, we should hastily conclude that everything of the sort is contraband. You cannot properly put it upon God, supposed real, to prove Himself, His Word, or any of its doctrines by any given species or form of argument, arbitrarily selected. We have no right to instance Ontology, or Physiology, or History, or Astronomy, and insist upon it that God shall prove Himself by means of our favorite science and under our favorite forms of reason. A God is Himself best judge of what arguments it will be best for us to have — assuming it best for us to have some — and He is entitled to choose His own. It would be quite as presumptuous for us to dictate to Him in this matter, as it would be to dictate to Him what experiments he must submit to for the increase of our faith. But because it would be improper for us to demand that God should prove Himself to us by certain arguments of a class chosen by ourselves, we do not conclude that all arguments for that object are unlawful. We may be authorized to desire arguments in favor of what we are called on to believe; if so, we are authorized to ask that they be sound and sufficient — only we are not allowed to require that they be of this or that sort, or that they come to us in this or that way. So with these experiments. We cannot appoint to God what arguments for Himself He shall allow us, nor can we appoint to Him what experiments He shall allow us. Nevertheless, there may be good and lawful arguing in that quarter to be done; and there may be equally good and lawful experimenting. There are direct practical trials of God and Scripture which we can make for the benefit of faith, which are no setting up of our own wisdom, no presumptuous dictations to Him who may prove to be the Most High, no familiar and irreverent applications of as it were hammer and acid to the Holy of Holies, to the ark of the covenant, and even to Him who sitteth between the cherubim. But they are such as Faraday and Brewster, reverent interpreters of nature, seemed to be making when from a distance some disciple watched them poring with shaded eyes and shrinking, half-retreating attitude over a beam of light fresh from the sun, or the keen elemental fire that leaps from the batteries of galvanism. And the doings may all be in the manner of yon uncovered and hushed physician. Is not that sick man of monarchs the greatest and best? Is he not the great warrior and statesman and father of his people; and does not his empire kiss at once sunrising and sunsetting, sweep the breadth of three continents, swelter under the golden suns of the Bosphorus and glisten in perpetual whiteness beneath the frozen pole? But now he is prostrate; and that medical adviser enters with bare brow and muffled step. He places his finger on that pulse as if rose and sank with it the majesty of a nation's life, and of a dynasty awful with the glory of a thousand years. In the same spirit may we and should we deal with these imperial questions relating to august God and Revelation.

The God and Revelation of Christendom have furnished their own practical tests. They have shown us what experiments they are willing to have us make on them. We are not to make arbitrary and unauthorized experiments; none whatever in a spirit of lightness or audacity; but such as are actually furnished in the Scriptures we may freely use, minding to do all with a modesty befitting the great conceptions with which we deal.

Among these lawful and actually furnished experiments are the following — which I offer, not in the name of practical religion, but in the name of Modern Science. The Scriptures make many, clear, and striking promises to liberality. Thus; "The liberal soul shall be made fat, and he that watereth shall be watered also himself. Honor the Lord with thy substance and with the first-fruits of all thine increase; so shall thy barns be filled with plenty, and thy presses shall burst out with new wine." "Give, and it shall be given you; good measure, pressed down, and shaken together and running over, shall men give into your bosom." And so on in wonderful profusion. Now, unbeliever or weak believer, make an experiment. Be liberal, and see whether these promises are not fulfilled to you. See whether the property, or what you are disposed to accept as its full equivalent, does not accumulate. Then you will have put to a direct practical test both God and the Scriptures — the reality of the one and the divinity of the other. — Again, it is written that if we pray for the Holy Spirit and religious blessings in general with sincerity and earnestness, they shall without fail be given. For blessings of this sort the language is, "Ask, and it shall be given you; seek, and ye shall find; knock, and it shall be opened to you: for every one that asketh receiveth, and he that seeketh findeth, and to him that knocketh it shall be opened." Now, unbeliever or weak believer, make an experiment. Perseveringly put your heart into prayer for these blessings, and see whether they do not come. So will you put the alleged revelation to a searching practical test, and, as it were, bring the reality of its God and of its inspiration within reach of the senses.—Again, it is written that they who follow conscience faithfully shall in so doing come to something better than the light of nature, namely, a written revelation—come to an assured faith in Jesus and His doctrine. "If any man will do His will, he shall know of the doctrine whether it be of God or whether I speak of myself." Now, unbeliever or weak believer, make an experiment. Go to walking most carefully according to the light you have on matters of duty, and see whether faith in the Scripture and a scriptural God does not shake a freer wing, and soar nearer the sun, than ever before. So will you bring religion out of the hands of Plato into the hands of Bacon; will transfer it from the dry world of tradition or logic into the green world of actual personal experiment; will, as it were, put it where your hands can feel it, and where, like the unbelieving apostle, you can even put your finger into the print of the nails, and thrust your hand into the side of both natural and revealed theology. Act on the Bible itself carefully as a rule of life, and see whether it does not most palpably agree with your constitution, as much so as delicious water and bread do with your body—so showing by a personal trial of your own that the two were made for each other by the author of both. You can safely make these practical trials. They are not of your selecting and dictating. They are furnished ready to your hand by the parties who are to be tested by them. These parties not only consent, but urgently request to be tested by them. If this fair offer and manifold urgency be successful — if you conduct the experiment with fitting reverence, gazing with shaded eyes, and stretching out trembling, half-retreating hand toward the possible Uncreated Light and Celestial Fire that condescend to offer themselves to the criticism of your experience — you can, so says the alleged and alleging Religion, if you are without faith, get it; if you have small faith, you can increase it; if your own faith is strong like sight and you wish to impart the like to the weak and the doubting and the disbelieving around you, you can powerfully say to them, "Sirs, the truth and excellency of the fundamental religious doctrine, of the Theism and the Christianity, are to me not mere matters of tradition or logic, but matters of direct personal experiment. I have, so to speak, 'tasted and seen' that God is, is what the Christian Scriptures represent Him, is the author of those Scriptures. Take my testimony, as you would if I should say that I have smitten on this wood and found it to be hard, or have put a drop of nitric acid on this metal and found it to be gold."

This method is strictly scientific. It is just as Baconian as the process that has built up our chemistry and our other natural sciences into such admirable splendor. It is the eldest-born of the Inductive Philosophy; and if any claim that its accent is that of the pulpit, I answer that it is equally that of the laboratory. In my opinion, any scheme for promoting an intellectual faith in God and the Scriptures that does not include this Experimental Method, is as much against the true modern philosophy as against religion. More than this — any scheme that does not place this method in the foreground, as having supreme rank and as vastly better than any argumentative method can possibly be by itself, is a failure. What is commonly called arguing, namely, the establishing and putting together certain propositions, and then drawing a conclusion from them, is, no doubt, a very useful thing — nowhere, as we have in due time to make evident, more useful than in the field of fundamental religious doctrine. At the same time it ought to be distinctly professed that in this field no possible argumentative proof can equal in some main respects its elder sister, the experimental ; and that no actual logic has equalled it in point of success. Such practical experiments as I have mentioned can readily be made by men of the narrowest leisure, capacity, and knowledge : their ordinary pursuits need not be interfered with in the slightest. Not so with a large portion of arguments on the same theme. To be properly estimated, these require talent, education, and studious leisure in no small degree. But who cannot put God and the Scriptures on the test of this actual experiment? Who so poor, so weak, so ill informed, so uncultured, so busy that he cannot try these things by his liberality, by his prayer, by his conscientious living? Such a variety of easy practical methods enables all the world to become critics in religion. And, in point of fact, many times as many converts from unbelief have been made by them as by all the exertions of logic. I do not as yet say that logic has any proper place within this field; but it has been widely supposed to have, and so has been sent out in vast masses and in every style of armament to conquer the unbeliefs and disbeliefs of the world. It has had its successes. Spolia opima have been won. Triumphs have been decreed. But never such triumphs as have been granted to the Experimental Method — triumphs of the first order — not ovations, but triumphs — triumphs in which laurels have waved like a forest, and in which chained champions and monarchs have gone in long procession after the captive wealth of empires and races. The great body of Christian believers in all ages have had no other rational faith than such as they obtained and maintained by actually putting the Christian Religion, with its God and inspired Bible, to the test of practice: thus verifying in their experience its adaptation to the human nature and condition, its transforming power, and the faithfulness of its promises. Moreover, an argumentative faith, as well as a traditional one, is observed to have always a certain deadness about it till it is supplemented and inspired by the faith that comes from direct personal experiment. The latter, when acquired, becomes a soul to the former. It paints its clayey cheek with speaking vermilion. It lights up its lack-luster eye with the beautiful fires of thought, and feeling, and force. Oh, how that poor mass of flesh and blood, by courtesy called man, and which yesterday could not stand on its feet or even scarcely fetch a breath as it lay with glassy eyes by the wayside — how strongly to-day heave the arches of its breast; how buoyantly it springs to its feet, and, with head uplift to heaven, plants itself like a pyramid; how swiftly now and strongly it marches hither and thither, with every feature alive, and every muscle strung for doing and daring! A soul has entered the clay. The form now tabernacles a power. Welcome, O great, beautiful, glorious Transformer! No vinous life art thou, no life galvanic, but true Divine Breath, the mighty afflatus of actual personal experience in religion: lo, thou hast wrought that wondrous change, and made of the mock man a real one! — The proof of God and the Scriptures by personal experiment has also this advantage over any possible proof by argument; namely, that it has an intrinsic value of its own, apart from its character as a means of faith. In general, an argument is worth nothing beyond its tendency to produce faith. But the course of beneficence, of prayer, of conscientious living, is in itself always a mighty blessing, even were no religious faith to result.

According to the Christian system of religion, everything depends on possessing faith. We must believe in God, in the Scriptures, and in their principal doctrines; and the broader and deeper our belief is, the better it will be for us. If we have a sincere faith, then we need to make it great; if it is great, we need to make it royal; if it is royal, we need to make it perfect; if we could say it is perfect in ourselves, we should still need to originate or improve it in a host of others as being the greatest favor we can confer upon them. So that to all of us this broad method, this scientific method, of faith by means of personal experiment and induction, is a matter of high moment. A plentiful use of it is the great want of the times. And we may be sure that quite too little account is made of it, even among most of those who have been most indebted to it for such measures of faith as they have. Even these too often assume that all improvement in this foundation grace must proceed in the way of argument. If themselves need to be stronger believers, they do not think of experimenting: it is either waiting for what the winds will bring them, or it is arguing. If others are to be rid of their doubts, they are, primarily and perhaps solely, to be argued with. Here is a profound mistake. What at the most is secondary, is made primary. It is not the reason that is so much at fault in cases of deficient faith: it is the practical part of us. The remedy ,s not so much syllogizing as it is doing. It is not argument and experiment that is wanted: at the most it is experiment and argument. The one is the lightning that unsolders and seams the masonry of Doubting Castle: the other is more like the billowy and thunderous air that rolls in afterward, wave upon wave, to help in shaking the ugly structure to pieces. The foremost great thing to be done for our weak-faithed selves and our weak-faithed neighbors is to send them to school in the first department of the Inductive Philosophy. They must be put up to that which in religion answers to the hammer of the geologist, the acid of the chemist, and the prism of the optician. They must personally try practical tests on God and Scripture. They must take such tests as the Christian Deity and Scriptures offer to be tried by, and faithfully and reverently go into a sacred experiment. This I have felt bound to put forward as the leading work to be done in favor of faith. Let these men of scant faith all around us try God and the Bible by their promises. Let them test these great allegations by generous beneficence, by hearty persevering prayer for spiritual blessings, by nonestly endeavoring to go by the obviously just rules of the Scriptures in all the every-day walks of life. This will do more for them than libraries of argument could do without it. It is a means universally accessible, has done wonders in its day, and is waiting at the gate of every man who needs more faith than he has, to do them again for his benefit. It may not do them at once ; it may render its proofs somewhat tardily and amid some discouragements; but it is according to experience that for any given man this method will accomplish a quicker as well as a stronger faith in God and the Scriptures than any other method by itself could have done for him. If the man is such in his natural turn of mind and habits that it will take years at the Experimental Method to convince him, he is such a man as would hardly be convinced by a lifetime at any other school. But the crowning thing is that the experimentalist is sure of great success in the end. Whatever the adverse appearances and long delays, the promise that he shall "know of the doctrine," will at last come to fulfillment. He shall not die till his faith lives. And though, in some rare instance, he should be tried with as much delay and as great seeming adversities as Joseph had while on his way to the fulfillment of his dreams and the premiership of Egypt, still, the faith which he shall surely reach at last shall be of that royal kind that will plenarily pay for all. As with the Hebrew, his De Profundis shall full surely become his In Excelsis.

To the pit-bottom he sank,
 That poor Hebrew lad,
And the thirsty darkness drank
 The light within him.

" Now all things go against me,"
 Said that poor sunk lad,
Earth-eaten, waiting to be
 Eaten of famine.

Up through the earth-pit dreary,
  Swung that poor sold lad
Into worse pit of slavery,
  Arab, Egyptian.

"Still all things go against me,"
  Said that poor slave lad,
As sun-scorched, thought-scorched, through sea
  Of sand he falters

To Misraim — to be bought,
  (Ah, poor chattel lad!)
And wrought with the lash for nought,
  Like soulless cattle.

O emir-sprung and petted,
  Now sunk, sold, slave lad!
How is thy poor heart fretted
  To cry, "Against me!"

"Against me! yes, against me!"
  Not so poor blind lad!
Where pain plies red beak on thee
  Thy kingdom enters.

Fell pit and master anoint
  Thee Pharaoh, lad!
Fell pit and master appoint
  Thee chief sheaf — star prince

To sun, moon, and brother stars,
  (O true dreamer lad!)
And brighter stars whose rays are bars,
  Ruling Osiris.

So judge not by the seeming,
  Faithward fighting heart!
The rod that leaves thee streaming,
  Will turn thy scepter.

If thou for true faith equipt,
  Meet pit and master,
It shall sure crown thy Egypt,
  Here and hereafter.

II.

ARGUMENTATIVE METHOD.

Ἕτοιμοι δὲ ἀεὶ πρὸς ἀπολογίαν.

Ἐκ τοιαύτης, ἄρα, Ἀρχῆς ἤρτηται ἡ φύσις.—*Aristotle.*

II. Argumentative Method.

SECOND LECTURE.

ARGUMENTATIVE METHOD.

I HAVE spoken of the Experimental Method of proving the Christian God and Scriptures — of ts nature, mode of use, strictly scientific character, and paramount place in a wise scheme of religious evidences.

We come now to the Argumentative Method. I ask your attention to remarks on its Possibility, its Propriety, and its Possible Profit.

The possibility of logically proving God and Scripture has sometimes been questioned on *a priori* grounds. On such grounds some persons have questioned the possibility of proving anything by argument — skeptics, who have doubted not only that anything can be proved, but that anything can be known, even the fact that we can know nothing. The critical philosophers, so called, with Kant at their head, without going so far as this, are still decided that there can be no argumentative proof of supersensible objects — that is, of objects not directly cognizable by the senses, such as God and religion — and of course no logical proof of the Christian Scriptures as being God's message. Still others, bearing such names as Fichte, Shelling, Hegel — the Anti-criticalists, Idealists, and Pantheists, especially of Germany and France — declare that there may be arguments to prove a God, but none to prove such a God as the Christian Scriptures teach, namely, a personal God external to the human mind and distinct from Nature. Well, a God who is a mere idea, or the moral order of the world, or the sum total of Nature, is no God at all to a truly English mind, and can issue no message.

It would be impracticable, in such a course of lectures as I propose, to examine the grounds on which these men rest their conclusions. Fortunately it is not necessary. If a man should deny the possibility of a good watch on abstract considerations, our best method of dealing with him would be to show him such a watch. If some Dr. Lardner should deny the possibility of crossing the Atlantic by steam, the most satisfactory reply possible would be to embark him in one of the hundred steamers plying between the two hemispheres, and actually transmit him to England by the impossible method. So the best way of dealing with such speculations as deny or doubt the possibility of good arguments for God and Scripture is actually to produce such arguments. This is the way in which the Baconians effectively answered the old philosophy. Said that philosophy, "A true science cannot be built up by experiment and induction: it must be done by reasoning from general intuitions, and," as some said, "other general truths forming the original furniture of the mind." This doctrine stood unfalteringly against ages of skillful dialectics. And it was not till the true philosophers turned from wasting time and strength in logically combating this position, to the task of actually building up the natural sciences in the way pronounced impossible, that those Platonists met their silencing refutation. What could a Ptolemist say, with his eye at one end of Galileo's tube and the phases of Venus at the other? What could any philosopher of the old stamp say, in the presence of the actual Astronomy or Chemistry; which, rooted in observation and experiment, had risen in the course of a few years, by mingled induction and mathematics, into such lofty and wide-branching majesty of stature and fruitfulness as the old system had for some thousands of years been always promising, and never even beginning to accomplish? There was no resisting the eloquence of such examples. Yes, experimental and inductive sciences doubtless can be, because they are; and so the Platonists amended their doctrine of the impossibility of such sciences into the doctrine that they are a less noble and fruitful kind of science than the German metaphysics. Let us try to walk in the steps of those fathers of the Inductive Philosophy. Let us attempt no answer to those who deny or doubt the possibility of good arguments for God and Scripture, save the actual presentation of such arguments. If from the beginning, and under the ablest hands, no such argument has ever been constructed, it would do little good at this late day to establish its abstract possibility ; if one such argument can be actually shown, all the cloudy speculation against its possibility will meet the most evident and signal annihilation possible.

Besides these professional metaphysicians — as they were for the most part — some eminent Christian theologians have denied the possibility of a logical basis for religion. Their ground has been twofold. Some have said that God and His written message are as plain facts as any of our first principles, and consequently, according to well-known law, can only be darkened by questioning and reasoning about their reality. Others state themselves in this manner. Reason in man is a shattered instrument in shattered circumstances. It is so shattered within and around that no reliance can be placed on its verdicts on fundamental religious questions. Look at that seething chaos of opinions and reasonings which from the earliest times has borne the proud name of philosophy, and in which many a great logician, "floating many a rood," has lain bewildered — the puerile conceits, the muddy obscurities, the gross contradictions and self-contradictions, the stark absurdities, the terrible heresies, on whose windy and yeasty bosom reputations and schools and systems have tossed, and collided, and gone to pieces! The adventurous voyager, "through the shock of fighting elements, on all sides round environed, wins his way; harder beset and more endangered than when Argo passed through Bosphorus, betwixt the justling rocks, or when Ulysses on the larboard shunned Charybdis, and by the other whirlpool steered." Behold what the boasted reason can do for the world — especially in radical discussions! See — its very name has fallen into contempt! Is such a guide to have our confidence? No! say these theologians emphatically; and they feel themselves confirmed in their strong negative by the manner in which the Scriptures speak of certain things called philosophies and wisdoms — declaring that the world by wisdom knows not God; that the faith of christians stands not in the wisdom of men, but by the power of God; that men may be spoiled by philosophy, and should avoid oppositions of science falsely so-called. Their conclusion is that the reason which some men deify is at best but a fetich — that the true guide within the field of fundamental religious doctrine is faith; meaning, not a belief in God and Scripture resting on logical evidence, but one independent of such evidence and supernaturally given to those to whom it is appointed, or who pray for it and honestly endeavor to follow conscience. This faith carries men to the Bible and to prayer for that guidance in opinions and practice which their dilapidated reason is not qualified to give.

Men professing such views have not been very numerous among Protestants. Once in a while, however, they make their appearance. And in almost all our communities there is, I imagine, some such vein of thought silently underlying a portion of the casual reflection on this subject. But the answer is easy. It is true that human reason is in a fallen state, that it gives no absolute demonstrations in questions not mathematical, that many of those who have worn its uniform and carried its banners have left a very humiliating history, and that some of even its most gifted sons have in its name played off most extravagant quixotism and errantry of speculation. But how does one know that this mortifying exhibition is not due, partly to the impracticable nature of some of the questions discussed, and partly to the improper method and spirit in which most of them were examined? Is not the cause adequate to account for the result? But these impugners of reason, as employed on the fundamental religious theory, have their positive refutation in the examples and precepts of the Book which they acknowledge as the final arbiter of every question on which it pronounces. The Christian apostles argued freely with men in behalf of both God and Christianity. Their habit was to go into the temples, the markets, the synagogues, and there argue for their cause with Gentile and Jew, with atheist and infidel. Especially was this the habit of that princely logician Paul, who, wherever he went, plied the sharp edge of his remorseless logic; now in caviling Jerusalem on the scholars of Gamaliel, and now in sneering Athens on the scholars of Epicurus and Zeno. Who instructed Christians in the midst of a Godless and Christless age to be always ready to give a reason of the hope that was in them — also to prove all things, holding fast that which is good? The doctrine of the first teachers of Christianity evidently was that there is both a need and a reliable way of employing reason for establishing the reality of God and His message, and that the egregious follies and blunders that sometimes occur in the course of the logical process are to be set down, not against reason itself, but against its mismanagement.

Among some who allow the possibility of an argumentative method, it is still a question whether such a method can properly be attempted in behalf of faith. Many plain Christians are of this class. They have a strong feeling against any logical religion. The sight of such a great body of it as some European libraries show — thousands of volumes from Plato downward, and displaying an amount of genius, culture, and research vastly more considerable than their number — such a sight would make on their minds an impression of prodigious waste, to say the least; waste of time, money, pains, faculty. They have never felt the need of such books. They are strong in faith — thanks to early training and the experimental method — without any help from such a quarter; and it is hard for them, with their very limited acquaintance with the nature and extent of the attacks made on Theism and Christianity, to realize that any persons can require such help, or be at all the better for it. Especially is their feeling strong against logical Theism. They say that the Scriptures assume the being of a God, and so should we; that at heart His reality is doubted by none, all show to the contrary notwithstanding; that, if there is any such thing as sincere atheism in the world, it uniformly began and solely rests in a bad state of the heart, and so will not be reached by any mere logic, however conclusive and abundant. The same things, mutatis mutandis, are alleged against logical Christianity, though with somewhat less emphasis and prominence.

Do the Scriptures assume a God, and their own binding authority as His message — at least so far as argument is concerned? In one sense, yes — in another sense, no. It is not necessary to an argument that it take the form of a syllogism, with its major and minor and formally drawn conclusion. It is enough that such facts and principles are placed before the mind as seem to authorize, and naturally lead the reason to make, the desired inference for itself. This much the Scriptures do — in behalf at once of both God and Revelation. They attempt to show in themselves prophecies, miracles, and supernatural adaptations of various kinds, from which, if real, both Theism and Christianity are directly inferable in one breath. In such informal logic as this they may be said to abound. Further, the Scriptures claim that Theism is sincerely rejected by "fools who say in their heart that there is no God" — also, that Christianity is sincerely rejected by such men as Paul, who "verily thought he ought to do many things contrary to the name of Jesus of Nazareth." Indeed, if any reliance can be placed on testimony and observation and the ordinary laws of evidence, the cases of real unbelief and even disbelief in a God, as well as in the Scriptures, are by no means few. Many say they doubt or disbelieve; they do it with all facial show of sincerity and self-knowledge; above all, they act as if they disbelieved. What better proof could we have? As to such doubt and disbelief, supposed real, always finding its origin and support solely in a bad state of the heart, this may be admitted without admitting the inability of logical religion. The guilty heart must operate to produce and sustain the atheism and the infidelity by perverting and blinding the intellect; and all the light and just impulses we give the intellect are so much natural opposition to this effect, and may even work backward toward reclaiming the guilty heart itself. Thus the oarsman works his way up the river against the current; thus some potent essence, or heat, or sound creeps backward through the atmosphere against the wind; thus summer, beginning at the lowest edge of the glacier, steals drippingly and destructively upward till it reaches and melts the very fount of the icy cataract and sows flowers and perfumes around it.

Among those who admit the propriety of argument in behalf of Theism and Christianity, there is great difference of opinion as to the amount and kind of advantage possible from it. The expectations of some are enormous : the argumentative method is both the "ποῦ στῶ" and the lever which can move the world. The expectations of others are exceedingly moderate; indeed, so very moderate that they hardly find sufficient motive to give any thorough attention to the believing logic, from whatever source it may come. I have thought it desirable at this stage to state my own views on this point — partly as a key to my method of treating my subject, and partly because I should deem it equally unfortunate for any of you to come to the actual arguments with expectations either extravagantly large or extravagantly small, as to the advantages that may accrue from them. In the one case you would be disappointed into discouragement and an undervaluing of such utilities as may be found belonging to the argumentative method; in the other you would enter on the subject with too little interest to give it proper treatment.

I am disposed to claim great utility for the argumentative method. But I do not suppose this utility to lie mainly in quarters where many would naturally first look for it. It does not lie mainly in its power by itself to convert atheists and infidels into believers. Nor can any stress be laid on its value as a means of weakening the unbelief of such persons in the way of disputation with them. We cannot even claim for it that it is the leading means of sustaining and strengthening faith in God and the Scriptures where such faith exists. We have large admissions to make against logical religion at all these points. It is found in experience that religion is seldom proved to the satisfaction of men by any merely logical argument whatever. When men become theists, they not only generally become such by a sort of proof that accredits to them Jesus and the Bible at the same time, but this comprehensive proof itself is generally something besides syllogisms, or what can be resolved into such. It is the proof by the experimental method. It is the proof by the experimental and argumentative methods combined and interleaved — that composite method, like the student's classic, whose alternate leaves of a richer texture than the rest and left blank for that purpose, give in his own hand his own personal thoughts and results ; that composite method, like the illuminated missal, whose every other leaf is pictured in silver and gold with the thoughts and feelings more dimly expressed in the neighboring words. The man feels that his wants are not met, that his nature is not fed, by infidelity and atheism. He knows it safe and reasonable and hopeful to renounce his sins. He begins to read and act on the Scriptures as being a practical system in at least general accord with his conscience. He thus finds his way to prayer to a possible God. And the result of all is that at length he discovers himself to be in possession of a measure of faith. Very likely he himself hardly knows how his mind has reached this point ; very likely he has attempted no formal study of Theistic and Christian evidences — nor even consciously given them any attention at all ; but in some way, certainly not purely argumentative nor even chiefly so, his difficulties and doubts have noiselessly thinned away like the fogs and chills from some morning landscape. It is in some such way as this that unbelievers usually become theists and christians. — And it is the great way, too, of preserving and increasing faith where it exists. The believer always intensifies himself far more by conscientious acting than by logical arguing. A day's careful discharge of duty will do far more to heighten his sense of the reality of God and of a Divine Scripture than will many a day's study of Paley, or any other writer on evidences. Instead of being the great means of producing, supporting, and increasing religious faith of any kind, mere argument deserves no notice in comparison with the easier and universally applicable practical method. And, further, we must confess that the mere argumentative method, when applied in the way of disputation, not only seldom removes, but generally strengthens unbelief in our opponent. Gladiators may conquer, but must not be expected to convince. A blow from a steel-glove rarely makes a man feel more amiably toward either the person or principles of his antagonist. The breaking of lances may be a very fine thing to lookers-on and the victorious champion: but it is a very uncomfortable and wrathful thing to the Templar, as he rolls in the dust amid the blare of trumpets and the swarming glances of tier upon tier of the valiant, the noble, and the fair. Will he ever feel kindly toward the Disinherited Knight or any of his belongings? Do not expect it. Rather expect to find him a more bitter Templar than ever. And disputation with lips, no less than with lances, whatever it may do for silent observers, may be expected to confirm our opponent in his views, by enlisting self-love and ambition and the passions of conflict in their support — leading him to give specially favorable attention to the plausibilities on his own side, and specially prejudiced and carping attention to the plausibilities on the other side.

These admissions must be made. But they are by no means an admission of the small utility of the argumentative method. Its uses are real and great, though not such in kind or degree as some claim. Granted that there are such things as sound scientific arguments in favor of God and the

Scriptures, there is a strong presumption, certainly, that they can be made to serve some very valuable purpose : and something of a presumption, too, that what so many great and good men — bearing such names as Newton, Locke, Clarke, Berkeley, Whately, Miller — have deemed greatly useful, not to say necessary, and on which they have expended such a wealth of toil and culture and genius as likens them to that Jupiter who is said to have once showered himself on the world in the form of gold, is far from being a vain thing. And the presumption should become a certainty to the christian when he finds that his Scriptures teach him, both by apostolic example and by precept, to be "ready always to give an answer to every man that asketh a reason for the hope that is in him." But multitudes of christians have very little faculty for suitably bringing up from the depths of their own minds the reasons for believing which they actually possess. They sit on the well; there is water enough in it to supply Jacob, his children, and his cattle; but they have nothing to draw with, and the well is deep. And it is very desirable that science and scholarship should come forward to put them into connection with their own abundant waters; so that they may pour them out freely at the curb-stone to refresh, not merely themselves, but the weary and thirsty men who are continually passing.

One use of the argumentative method is that it will serve in many cases to withstand the decay and fall of faith, especially in the young. Ingenious men have started numerous objections and woven numerous sophisms against the Christian Scriptures and their God. Many of these are well adapted to perplex and deceive the young and incautious mind. They are perpetually turning up, covertly or openly, in books, magazines, newspapers, popular lectures, conversation. Almost every community, even in New England, has some one or more, who, to the extent of their influence, are confessed perverters of the opinions of the young; and pride themselves on retailing wherever opportunity offers, the sneers and arguments of prominent infidels and atheists. No guardians, however careful, can prevent their wards, as they come forward in life, from meeting with these anthropophagi. And it is very desirable that what cannot be prevented, should be prepared for; that the faith which tradition, aided by instincts and casual observation and a certain unconscious logic, has already established in multitudes of the young, should be fortified in advance with well-considered grounds of reason against the sophistries they will have to encounter; certainly, that there should be within their reach at the time of danger the natural antidote to the poisonous error in the shape of its logical refutation. Of course the precautionary instruction is the best. And here that great body of logical religion which scholars have carefully digested and published to the world, comprising reasonings of all sorts and in all the moods and tenses of thought and expression, will serve a most valuable purpose. The Mentor of Telemachus can go to this roomy Panoplon of all the Greeks, and obtain from its endless variety just the argument adapted to the capacity and way of thinking peculiar to his ward. And it will be received with great freedom and held with great pertinacity; for, as yet, the young man is a believer. The consequence will be that when in course of life he falls in with the cavils and sophistries of unbelief, however ingenious, his mind will suffer no perplexity and his faith receive no shock. It will not become a leaning tower of Pisa. He will not be the soldier brought to his knee by severe wounds and loss of blood. His friends will have the satisfaction of seeing the assailing darts, however deftly and forcibly flung, rebound harmlessly from the armor of proof provided in anticipation of such attacks. Without such forearming they would have seen him, not only in great risk, but actually wounded, prostrate, and dead.

Moreover, it may properly be claimed that the argumentative method will almost uniformly do something to strengthen the Theism and Christianity of practical believers who will give it suitable attention, especially those of the more intellectual cast; and such will be likely to give it attention. No christian, however brawny his faith, can say that it is as strong as it is desirable it should be, and as it might be. He has merely a good beginning of what admits and calls for indefinite improvement. The more nearly his faith likens itself to sight, and the Christian God and Revelation stand forth to his mind as do the oceans and mountains and stars, in their massive and inexorable reality, the purer will be the heart he will bear and the life he will lead. How shall his faith receive this needed enlargement? I have repeatedly spoken of the great method, that practical method, in comparison with which no other deserves a thought. But there is another, of considerable independent value in its place; that of familiarizing the mind with that wide variety of logic in behalf of the fundamental religion on which have been expended so much of the best thinking and expression of the world. If one is already a believer, there is nothing to prevent this sound argument from taking its natural effect upon him; he is predisposed to welcome it, and to give it due weight. Under these circumstances, especially if his mind is of the more thoughtful and investigating character, he will find his study of the logical evidences giving his faith new outspread and foundation. The Thesaurus of logical religion has become an exceeding great Nineveh, of three days' journey — has become hundred-gated Thebes, able to send forth a myriad warriors from each gate. One is sure to find, somewhere within its wide precincts and amid its metropolitan resources what is suited to his peculiarity of habit as a thinker and as a christian. Who has the freedom of the national Commissariat will be sure to find, among the prodigious stores of necessaries and luxuries that crowd its roomy depots, something to suit his peculiarity of appetite and constitution; who has the freedom of the national Mint, where are piled up, in glittering stacks, tons of coins of every precious metal and every denomination, can surely find both change and capital enough for any personal expense or reasonable business crisis that has come upon him; who has the freedom of the national Arsenal, and looks around on the weapons offensive and defensive, ancient and modern, foreign and domestic, for siege and battle, for land and sea, for officer and private, whose burnished steel and brass — not to say silver and gold — mix their terrible sheen from floor to ceiling, will surely be able to generously accommodate his own idiosyncrasies of enemy and campaign and strength and stature and skill, whatever these may be.

It can also be said of the argumentative method that, by itself, it may often weaken and occasionally overthrow atheism and infidelity. I say occasionally. Observation seems to show that, while the great experimental method must be chiefly relied on to do this work, now and then a case of conversion to intellectual Theism and Christianity occurs under the mere pressure of argument. Such were the cases of Galen, Thorpe, and Nelson; and the latter, in his "Cause and Cure of Infidelity," gives several instances additional. It is well known that the almost universal unbelief in Yale College at the beginning of the present century was completely overturned by the reasonings of its eloquent president. So long as the unbeliever is disputatious, so long as the spirit of prejudice and rancor is active, the soundest and most victorious of arguments will not take effect on him: but there are certain opportune and critical moments, certain Thermopylæ-passages in his life, when conscience and Providence have spurred up the mind to some measure of candid thoughtfulness; and, occasionally, at such times the religious logic succeeds in getting such a firm hold of the roots of unbelief as enables it to dislodge the evil upas finally from the mind. It does not take many such achievements as this to pay for all the labor that has been expended in rearing and equipping the argumentative method.

These several uses will be served by that method considered as an independent agency. But its great use is rendered, not as an independent agency, but as an auxiliary to the practical method. It is true that in order to the success, in a very considerable degree, of this primary method, not a single formal argument in behalf of God and Scripture needs to be constructed. Every man is already, informally, in possession of as much light from that quarter as is necessary to the successful working of the test by experiment. At the same time the operation of this method will be greatly facilitated, and carried forward to much larger degrees of success than it could otherwise reach, if combined with a patient attention to those arguments in which many of the ablest thinkers of the world have given the most apt and forcible expression to the rational grounds of faith. Men generally need to be stimulated to the faithful and persevering use of the experimental method. They are very reluctant, especially atheists, to put themselves on a strict course of conscientious living. But an increase of their suspicion that they are in error will help them toward overcoming this reluctance ; and this increase, as we have seen, a just consideration of the ample logic is likely to give — a logic already ample, but which may be made as much ampler as the strata of Geology are ampler than your geological cabinet. In the case of the atheist such just consideration will, in general, only be obtained in part and with difficulty. But, if his well-wishers watch their opportunity, they can find some time when the spirit of prejudice and cavil is sufficiently inactive in him to allow of his looking at the Theistic argument with enough candor to greatly increase his uneasiness and latent Theistic suspicions. And this will be so much increase of pressure toward that practical method with the aid of which, in all probability, his atheism must ultimately be overthrown. Judiciously handled, our logical religion may be made the great dynamical feeder to that experimental method which is the world's main reliance for faith. It is worth far more in this capacity than as an independent agent. It will serve religion much better by recruiting forces for another general than by attempting to lead them itself.

The argumentative may also minister to the experimental method in another way. Besides furnishing stimulus to use that method, it furnishes a better measure of the material used in working it. The conscientious acting goes to remove prejudice, balance the judgment, rectify the purpose, suggest love of the truth, and bring Divine assistance ; and thus prepares the mind to take just and clear views of certain facts and principles which are the rational grounds of faith. A certain amount of these facts and principles must be had under even the experimental method ; and this amount will get supplied in connection with it without any conscious investigation. But it is desirable to have as large an amount as possible : because the magnificence of the faith, if not its existence, depends on the extent of the material as well as on its quality. A tithe of the shapely blocks of white marble that make up the cathedral of Milan would make a very solid and beautiful structure ; but still nothing to compare with that august temple whose pinnacled and massive amplitudes now bear up three thousand statues to gaze across the pictured plains of Lombardy, up the white slopes of the everlasting Alps. By means of the argumentative method, ministering to the experimental abundant material, every one may have a templed faith like the Duomo of Milan. Whoever faithfully uses the method by experiment shall surely have a solid and beautiful sanctuary: but whoever, in addition to this, takes pains to put into the hands of this first of builders such precious and profuse material as the argumentative method can quarry and hew from out its vast Paros and Carrara, shall have a metropolitan temple for his faith, a Te Deum in stone to which angels shall delight to become pilgrims; within whose mountain of marble and beneath whose dome sweeping grandly heavenward, he shall find all climates equalized, and a secure and joyful home as long as he lives.

III.

APPLICATION
OF THE
ARGUMENTATIVE METHOD.

Καὶ δεῦτε διελεγχθῶμεν.

Σὲ τὸν Αὐτοφυῆ, τὸν πάντων φύσιν ἐμπλέξανθ'.

Euripides.

III. Application of the Argumentative Method.

THIRD LECTURE.

APPLICATION OF THE ARGUMENTATIVE METHOD.

IN the last Lecture I called your attention to the Argumentative Method of proving the Christian God and Scriptures — to its possibility, propriety, and possible profit. I now propose to begin the application of the method.

Its successful application depends on the recognition of certain principles, which, however plain and however generally acted on in other fields of moral inquiry, are very largely treated with neglect in this whole religious field on which we are now entering. I shall therefore devote a small space to their consideration.

What purports to be a moral truth presents itself at our gate, and asks for admission. Of course we have a right to ask for credentials. What sort and degree of credentials ought we to be satisfied with — at least so far as to grant the admission ? In the first place, it is very plain that if we proceed to demand anything of the nature of mathematical demonstration, a demonstration involving the impossibility of the opposite of the thing demonstrated, we shall demand too much. Proof of this kind is not possible in moral fields. We have not a single moral conviction that rests on such evidence, and never will have. We are now dealing with a class of ideas contradistinguished from those of quantity. And yet almost every man who holds out against a God — as well indeed as almost every man who holds out against Christianity, or who, admitting Christianity, holds out against any of the doctrines commonly ascribed to it, or who, admitting these doctrines, holds out against any of the duties it is commonly supposed to enjoin — will insist on having it proved to him, not that he is probably in the wrong, but that it is impossible he is in the right. "Prove to me," he says, "that antitheism cannot be true." "Prove to me," he says, "that antichristianity is necessarily false." "You say this is my duty: prove now," says he, "that the contrary is impossible in the nature of things." The demand is preposterous. No moral truth can have mathematical credentials.

Moreover, it is very plain that if we require in behalf of such truth evidence that carries with it moral certainty, we require altogether too much. Not that such evidence is impossible or undesirable within this field. Still it is too much for us to require as the condition of believing. Has any master of sentences, any standard of the art of reasoning, laid it down as a maxim that we are at liberty to refuse belief whenever we can avoid believing? Did Newton, or Locke, or any other honest great thinker since the world began, carry on his investigations of a moral kind under such a rule? Does any one do it — save when he seems in danger of finding an unpalatable truth? Is this the rule men carry with them into their politics and their business — resolutely refusing faith in anything till they have been allowed to put their fingers into the print of the nails, and to thrust their hand into its side? By no means. Their politics and business would hastily come to an end if they did; and their whole neighborhood would sneer at the impracticable men who are forever insisting on moral certainties and demonstrations, and will yield assent to nothing till absolutely compelled by Hercules and his club — that is to say, by an overpowering stress of argument. And yet almost every man who holds out against a God, or against the Christian Scriptures as His message — as well indeed as almost every man who, admitting these, holds out against any of the unpalatable doctrines or duties commonly ascribed to them — will insist on its being proved to him, if not that it is impossible he is in the right, at least that it is certain he is in the wrong. When reminded that there are no mathematics in any part of the moral field, he feels entitled to remember that there are moral certainties. These are what he wants. "Prove to me," he says, "that antitheism is surely false." "Prove to me," he says, "that the Bible is surely true." "You say," says he, "that this is a scriptural doctrine, duty: prove it beyond a doubt, and I will accept it as such." This man, perhaps, is not to be blamed for desiring evidence of the most convincing kind; his fault is that he must have this or none — that he will only begin to believe at the point where he should end, where faith, full-grown and fledged like an angel, is in the act of becoming sight. It would, undoubtedly, have been very pleasant to the man who, for a mere trifle, had just purchased an immense property in one of our Southern States, if he could have had, in addition to the deed of the recent owner, the fairly engrossed and broad-sealed deed of the United States of America, flanked by a certain constitutional amendment. But as he could not have this, he was glad to take up with a great deal less. He paid his half of one per cent. on the value of that Chatsworth, and joyfully took possession with nothing but that private deed in his hand — hoping in time to have something better.

Further, it is plain that if we require for the admission of a moral truth anything more than a preponderance of evidence, we require too much. What amount of evidence would be pleasing is one thing: what amount puts us under obligation to believe is another. Just as soon as, upon honest inquiry, there appear more probabilities for than against, then the foundation and obligation of faith are laid. We have no right to delay believing one single moment. No matter how small the apparent balance of likelihood is — though the equipoise of the scale is disturbed by only a single grain — we must yield our assent just as truly as though that grain were a mountain. We are not, indeed, bound to exercise the strongest kind of faith on such a basis; but real faith, proportioned to the balance of probability, we are bound to exercise. This is the indisputable and undisputed scientific law of reasoning — statute and common law. Logic is bottomed on this. It is both the soil that feeds its root and the air that waves its branches. It is that which men universally act on in affairs of business and all secular life. It is what we must act on in our religious inquiries, if we would treat religion and the mental laws fairly. When a man declares that he does not regard Theism and Christianity as sufficiently substantiated, I say to him, "What is it you mean? Do you mean that they do not fairly bristle with impossibilities of the opposite, like the Principia of Newton and the Mécanique Céleste of La Place?" "Oh no," perhaps he replies, "I do not suppose religion to be a science of magnitude, and that souls operate and moral ideas stand related according to the laws of quantity." "Do you mean that they do not stand forth to view and assent like the solar orb in a cloudless day, so that none but the stone-blind will fail to see the glory?" "Oh no," perhaps he answers. "I am not ignorant that blank certainties are the exception under the present scheme of life, that they properly end the faith rather than begin it, that to make them its indispensable conditions would ruin my present life, and so might ruin my next, if there is such. No, I am not so unreasonable as that. But this much I do mean: I must have a broad, heaped, mass of evidence; the scale on the side of God and the Bible must come down with a rapid and decisive stroke; my judgment must not be embarrassed with a large array of counter-plausibilities. Is not this reasonable?" It would be reasonable for you to be glad should moral truth happen to come to you with such heavy and shining credentials — broad-shouldered as an Atlas, and able on occasion to bear up the very heavens; but to say that it must come thus or come in vain is playing the tyrant with the first principles of a rational logic. You may ask a preponderance of probabilities in favor of God and His message as a prerequisite to faith: this may be your due from scientific religion. But if you insist on a jot more, you are unreasonable. And yet almost every man who holds out against God and His message will insist on having it proved to him, if not that it is impossible that he is in the right, if not that it is certain that he is in the wrong, at least that he is in the wrong by a manifold and overawing balance of probability. "Prove," he says, "that the plausibilities of the atheist and the infidel, though accumulated and spread out to the utmost, can be covered and buried fathoms deep by the plausibilities of the theist and the christian — that while the idea of No-God and No-Revelation has merely a hand-breadth of base, that which underlies our current religion stretches away over the whole rocky foundations of an empire — and I will believe." It can be done; but shall one who knows what scientific logic means presume to demand so much as the condition of believing?

Such are the principles with which one ought to approach the application of the argumentative method. I have asked you to recollect them — not because I wish to make the most of a little evidence, but because I wish to make the most of a great deal; or, rather, because I wish you to do simple justice to those Alps and Andes of evidence which, almost uncounteracted, have, in connection with the experimental method, bowed to the simple yet majestic faith of children such minds as Boyle, and Locke, and Newton.

OUR FIRST CONCERN IS WITH THE DOCTRINE OF GOD. AND, BY THE TERM GOD, LET US MEAN SIMPLY AN ETERNAL BEING POSSESSING POWER AND INTELLIGENCE BEYOND ALL CONCEPTION GREATER THAN THE HUMAN.

Such a Being I affirm to exist. At present nothing is claimed about His unity, or character, or government. Nor is it claimed that His power and knowledge are absolutely infinite; only that they are practically shoreless to our thought. This narrowness of thesis, while it simplifies the discussion to be undertaken, sacrifices nothing of result. The man who gets convinced that there is an Eternal Person who towers above men in might and wisdom further than thought itself can soar, passes easily forward to a conviction of the Divine unity, the Divine goodness, the Divine government, and the strict illimitability of all the Divine attributes. Enough momentum is acquired in going so far to carry him much further. Really the battle is gained for the entire Natural Theology. No intelligent man of these days and countries would think of making a stand at any other point after this keep of his castle has been yielded. That high central tower commands all the outworks. You can sling a stone from it into every square foot of the fortress. This is instinctively felt by the broad intelligence of the nineteenth century. Accordingly, there is not a theist in all Christendom who believes in more than one God, or in a wicked God, or in an ungoverning God, or in One whose natural attributes are not substantially infinite. Whoever believes in Him at all, confesses Him to be the one infinitely great and good Author and Ruler of Nature. It was not always so. There was a time when theists were pluralists. There was a time when men believed in Ahriman — nay, in both Ahriman and Ormuzd. There was a time when men supposed that God wrapped Himself in His august infinity, and stood contemptuously aloof from all the affairs of men. But that time has long since passed. Epicurus is dead. The Magians, and Manichees, and Gnostics are all dead. To attack their opinions is to attack corpses. To prove to a theist of this late day that God is one, or good, or infinite, or sceptered, is lost labor — save as it freshens an old truth. The man admits it already. He is, at least, "a modern deist." Whatever practical ignoring of the leading Divine attributes as taught in the Scriptures he may display, they are fully admitted theoretically. So the task before us is simple. All we have to do is to show that there is an Eternal Person whose wisdom and power are indefinitely greater than the human. Having this, as human thinking now stands, we have all — we have the unity, the goodness, the infinity, and the government of God. Still it may be necessary to notice some objections to these, as being in effect objections to the Divine existence.

At the outset it is plain that God is intrinsically possible. Personal beings are common objects: so that if there is any insuperable intrinsic difficulty in the way of the existence of a God, it lies in the attributes of eternity and comparative infinitude of power and knowledge which are ascribed to Him. But at least one eternal and absolutely infinite thing is known to exist, namely, space; and there is no more difficulty in conceiving of an eternal and infinite Person as being actual than of eternal and infinite space as being so. Also, if there is no God, there must be eternal matter with its eternal laws —just as hard a conception as an eternal mind. Also, on looking on one side of us down the long line of animated nature, we find it occupied with beings in perpetually descending and mutually approximating types till we come to such as are infinitesimally small and rude — mere monads trembling on the border land contested between the organic and the inorganic, between something and nothing: shall any say it is impossible that the line extends on the other side of us upward and away among perpetually ascending and mutually receding types of being, till, at last, by one prodigious leap, the geometrical series ends in a Being inconceivably great and glorious? Who has the right to say that, of necessity, himself is the last term of the series, or even the middle of it — that it does not go on expanding above him like an inverted pyramid of Cheops till the base of all is reached in infinite God and heaven? What if some rooted gelatinous polyp should assume to pronounce in this manner — as it looks around the mud-hole where it stands *facile princeps*, and as it follows downward with its nascent vision the graded life that swarms through its sphere till it reaches that infusorial mote which the microscope magnifying sixteen millions of times has only just brought to light — what if that polyp should assume to pronounce in this manner?

Next, I proceed to say that the God who is intrinsically possible is on the whole probable. This, according to the logical principles just stated, means that whatever objections to the Divine existence may be found are outweighed by the arguments for it; perhaps means that the arguments in the one scale are zero, while those in the other are the entire multiplication table. Let us see.

By far the greater part of atheists do not claim that there is any positive evidence against a God; they only maintain the insufficiency of the evidence for Him. Their attitude is that of doubters, not disbelievers. Nor am I able to find that any objections deserving of notice, besides the three following, are ever alleged or felt against the Divine existence, as admitted to be intrinsically possible.

The objections are these.

First, The miseries and moral disorders of the world, together with such natural objects as go to promote these.

Second, The absence of all overpowering manifestation of God in Nature and the government of the world: or, at least, the absence of an irresistibly universal faith in Him. "If there were a God," says or feels the objector, "He would so clearly manifest Himself, or otherwise summon faith, as to make doubt of His existence universally impossible. But, instead of this, all is silence, invisibility, and undemonstrativeness on the part of any such Being; and while some disbelieve His existence, more doubt it, and great multitudes have a faith troublesomely weak and unimpressive."

Third, The alleged fact that all things which need to be accounted for can be accounted for as surely and well by purely natural principles as on the supposition of a God; in which case we are positively required by reason and all scientific usage to ascribe the facts to Nature rather than to the supernatural as their probable cause.

Here we have three objections. The last objection, however, should be thrown out for the present. It really lies not against the existence of a God — at the most only against a certain class of evidences in His favor. What it means is that certain material atoms, with their properties and laws, will just as well explain the existence of, say natural organisms, as will the hypothesis of a God. In another place I shall formally deny this. At present I have only to point out to you that were the alleged fact incontestable, it would not lie against the existence of a God — at the most, only against a certain class of evidences in His favor, namely, that from natural organisms. Allowing that these organisms can be produced with perfect ease by the economies wrapped up in certain natural elements, it follows, if you please, that organic Nature cannot be appealed to as direct proof of the Divine existence; but it does not follow that there is no other proof to which we can successfully appeal — does not follow, either surely or probably, that God does not exist, or even that He did not actually produce Nature in all its glorious outspread. You, with your young muscles and hearts, are perfectly competent to ascend Mont Blanc, and place your feet on the very crown of that Alpine monarch; but this fact does not even make it probable that you were ever in his neighborhood even. You have never set eyes on his mighty slopes. You have never even dreamed of doing so. And even if it could be proved that at some time you have really done feats fully equal to scaling that snowy miracle — have really ascended mountains as arduous — this would have no tendency to prove that you have ever struggled up those formidable Savoyan steeps. Even so, were certain natural elements quite competent to produce the noblest organic wonders that ever took the name of solar system or of man, it would be no probability that they were actually produced by these elements. But suppose it were — suppose it abundantly proved not only that certain material elements are competent to organize Nature as we find it organized, but that they actually did thus organize it — what then? Does it follow that there is no God? At most, it only follows that the organisms of Nature are not available as proof of Him. We are cut off from a certain class of evidences that have been much relied on: that is all. Other evidences may exist. What hinders that a God should make one of the coeternities of Nature; and, though not the author of its organisms, nor even of the primal elements from which they proceed, stand among them and over them from everlasting to everlasting as absolute sovereign? Nay, what hinders Him from being the author of those very material elements whose wondrous properties for combination and organization have naturally peopled the heavens with sidereal systems and the earth with the glories of vegetable and animal life? Absolutely nothing. We are perfectly free to suppose that the whole verdant tree of Nature roots itself ultimately in God — that the famous questions of the origin of species and spontaneous generation, of which unbelief in these days is trying to make so much, are really but questions as to modes and times of a Divine operation. Does God organize Nature with His own hand through all these years and countries and spaces, or did He, vast periods agone, launch into being certain atoms dowered with all those subtle affinities and laws which in process of time would of themselves issue naturally in all the wondrous mechanisms of nature — behold here the true dilemma with which the Darwins and the Lamarcks threaten us! This the chief of them profess. They profess that their views are perfectly consistent with Theism. They shoot not a single arrow anywhere in the direction of a God. Every shaft flies exactly a quadrant away — neither for nor against. Grant them all they ask, and it still remains perfectly open to proof that a God exists, and even that He created and governs the whole august total of Nature.

Setting aside, therefore, the last of the three objections, as having no claim to be considered at this part of our discussion, however much it may have at another part, let us revert to the first objection, that from the miseries and moral disorders of the world.

Now, in regard to this objection, it ought to be plain that, if it has any validity, it is not against the existence of such a God as I now affirm, namely, an Eternal Being of power and intelligence inconceivably beyond the human. At the most, it is only valid against a good God. A state of the world checkered by sin and sorrow and deformity, is surely not inconsistent with the existence of a wicked Deity. It would not be out of character for such a being to neglect us, to afflict us, to abuse us to any extent or in any manner. Were the world one vast torture-house and pandemonium, it would still agree perfectly well with the presidency of one who hates, or cares not for the holiness and happiness of his creatures. Looking around the dungeons of the Inquisition has no tendency to draw into doubt the reality of the Inquisitor-General, whatever conclusions it may warrant as to his sweetness and mercifulness. Looking around on the *débris* of worn and crushed geologic periods never induces geologists to think of calling in question the presence among them of some enormous force: they only are put upon considering whether that force is Plutonian or Neptunian.

This is my first answer to the objection from the sins and sorrows and other maculæ observable in Nature. If it has any force at all, it is, at the most, only against the goodness of God, not against His existence. But really it has no force even against His goodness. God may not only exist, but clothe Himself with goodness as the sun does itself with rays, notwithstanding the earth is confessedly scorched and scarred with physical and moral evil. I wish to show this for several reasons. It is well to push the objection which has been so great a trial to many still further from our thesis — so to speak, out of sight of it as well as out of hearing — and, as it were, make assurance of its invalidity doubly sure. Does the son content himself with merely turning off by the smallest possible angle the arrow aimed at his sire? Does he not rather with forceful and indignant blow smite it a whole semicircle away?

It may also be well to show the invalidity of the objection as against Divine goodness, in order to forestall a prejudice against accepting any God that naturally arises from supposing, or at least fearing, that the God, when accepted, will have to be admitted to be a bad one. We all had rather have no God than one destitute of goodness; and this feeling naturally stands in the way of the reception of any logic, however conclusive, in behalf of a

God which may have this enormous want. Another reason, perhaps the most important of all. There are many to whom it seems that an Eternal Being of inconceivably great intelligence and power logically implies a good God and abundant evidences of Him, and that, consequently, any objection valid against His goodness is really valid against His existence. For the sake of such persons also — some of them believers of the choicest kind — I desire to go further, and show that the various evils, natural and moral, of the world are not against even the Divine goodness; are not, under the circumstances of the case, even the smallest presumption on the whole that among the existences of the universe there is not One whose eternal years of might and wisdom are auroral with the glories of a perfect virtue.

Notice the following things. First, if God were not strictly almighty, the limitation of His power would sufficiently account for the evil observable about us; we should be quite at liberty to suppose Him perfectly good. Second, if He were not strictly omniscient, the limitation of His knowledge would sufficiently account for the evil around us; and we should be quite at liberty to suppose Him perfectly good still. Third, if these two limitations were existing together — and our thesis does not assume the contrary — they would furnish us with double the explanation required to meet the objection without giving up one jot from a perfect Divine goodness. By giving up either the strict almightiness or the strict omniscience, we can surely save the goodness in all its entirety: by giving up both, we can double, so to speak, the assurance of our position. For my part, if compelled to choose, I should prefer to allow that God is not quite metaphysically almighty, or all-wise, or even neither; that although powerful and intelligent beyond all human standard and thought, better equipped in these respects than Zeus or Brahma was ever fabled to be, His oceans of might and knowledge fall somewhat short of being absolutely shoreless. But this sacrifice is not necessary. A perfect Divine goodness can be saved without it. And it seems to me not hard to do it — *especially in view of the peculiar nature of virtue, and of the manifest fitness of an outward condition of imperfection and sorrow to a race of sinners.* I ask you to emphasize this last thought. Let it be the background on which you project such facts as the following — not for the purpose of exaggerating them, but for the purpose of setting them forth in all the truthfulness of nature.

Notice what the aspect of the world really is. We do not see exclusively sorrows, and sins, and shadows. By no means. We see, besides, a vast deal of enjoyment — from mere comfort to rapture; from the obvious gayety of the mote in his sunbeam, up the long line of gamboling and singing and smiling Nature, with its hundreds of thousands of known species, to the mighty joy of a man who at least thinks he has gained the prize of eternal life. In addition, we see an incalculable amount of things fitted to give enjoyment—useful things, delicious things, beautiful things, sublime things; things grateful to the touch, to the taste, to the smell, to the ear, to the sight, to the soul; pleasant lights and shadows; sweet perfumes and sounds; golden grains and fruits; lovely features, forms, flowers, gems, landscapes, motions; glorious rivers and cataracts and mountains and oceans and skies—in thronging hosts which no arithmetic can compute. Further, mixed up with this natural good is a great amount of such as is of a still higher nature. No one is warranted in saying or believing that there is a particle of sin in any of the animal races below man. But there are many fair and noble spiritual qualities revealing themselves in numberless ways through these humbler but wide domains—fair instincts, affections, gratitudes; noble endurance, courage, skill. And altogether, within historic and our daily observation, there are—generously sown through the world like star-dust, and lighting up our atmosphere with all manner of lights, from the atomic phosphorescence of the fire-fly to the gayest November star-rain—comely orders and proprieties, generous impulses, charming amiabilities, graceful affections; beauteous industries, usefulnesses, purities, aspirations, hopes; exalted patiences, fortitudes, heroisms, loves, magnanimities, moralities, consciences—above all, pure solid Christian virtue in very many incontestable and even glorious instances, the record of which thrills us as we read; also, in the case of every human being, capabilities of a virtue of the most magnificent description, and far loftier than any that ever actually pictured and glorified the historic page. Further, it is observed that virtue has in its favor the suffrages of all consciences, and, confessedly, the general current of natural laws and events.

Now, this I say, that if you hold God responsible for the sorrows, moral disorders, and other disadvantages of the world, it is but fair to give Him credit for the happiness and virtue, and manifold advantages of all sorts, that exist. If you debit Him with those dark things, you should credit Him with these bright things. If the one class of facts is allowed to argue against a good God, then the other class must be allowed to argue in His favor. And it is simply a question which party argues loudest—the Red Roses or the White, the Guelphs or the Ghibellines, the noes or the ayes. Who is warranted in pronouncing that the noes have it? My ears have not discovered it, nor have yours, nor yours; least of all—those of the objecting atheist. Confessedly, the happiness of the world is far greater than its sorrow: almost every living creature has a thousand moments of comfort to one moment of pain. Existence, as it is, is almost universally considered a blessing, and so much of a blessing that not one in a thousand but would a thousand times prefer living on, with his average lot as to happiness, to being dismissed into annihilation painlessly, or even by way of paradise. Confessedly, the noxious things, the deformed things, the things that wound the senses and the æsthetical nature, bear no sensible proportion to the useful, the comely, the gratifying things that be-green and be-blossom this beautiful world. Let every man look about and judge for himself. Atheists not only confess, but profess it. They are forward to claim great things for Nature: she is to them the one worshipful Alma Mater: they practically deify her and her laws. Confessedly, there are through the multitudinous races below man more orders than disorders, more proprieties than improprieties, more things that are comely and useful in disposition and instinct and habit than there are things of apparently the opposite character. I suppose no naturalist of standing, whatever his religious views, would for one moment think of calling this in question. An open profession of it, on the contrary, in terms enthusiastic and almost poetical, distinguishes the chieftains of natural history. It is true that when we come to man — if we take the Bible-microscope and the Bible-micrometer for inspecting and judging the hearts of men, and not otherwise — we find more sin than holiness; but then we find by the side of what goodness does exist, and assisting most heavily to bear down its scale, this more than fairly offsetting great fact, namely, that the general constitution of Nature, and all human consciences without exception the world over, are founded and immovably continued from age to age in the interests of virtue. I say this more than fairly offsetting fact, especially in view of the essentially free nature of virtue. But from the stand-point of the objecting atheists the case is still clearer. These are the men who have never accepted the Christian view of the corruption of human nature, nor the Christian view of the nature of virtue. These are the men who have constituted themselves professors of the dignity of human nature and of the innocence of childhood — men with whom every amiable instinct and graceful propriety and pleasing amenity passes for solid holiness — or rather, men with most of whom there is no such thing as sin, only misfortune or contrariety to public opinion; that is to say, no sin but pain, and no holiness but pleasure. According to these views, the world is just as fair morally as it is physically and in its relation to happiness.

This, then, is the state of the case, especially according to the objector's own showing: on the one side much, on the other side more — on the one side ten suffrages, on the other ten thousand — on the one side a good God negatived by a chorus of tears and sighs from the night, on the other affirmed by a much grander chorus of smiles and songs from the day. What right has any man to favor the vanquished night-side of Nature; and record judgment, not only in defiance of charity, but in defiance of the logic of testimony? What right has he to balance the books against a good God, when really there is a large balance to His credit, according to the observation of all discerning men? He has none, and stands by the side of the man who hearkens more to the spots on the sun than to the sun itself.

Now, suppose a mind brought to this stage should suddenly become clairvoyant as to the future of this world, and discover a little in advance a golden age unfolding itself in every land and among every race of creatures — the new reign of Saturn, the sabbath of geologic periods, the tenth avatar of Brahma, and the millennium of Christ — say, if you please, a thousand years whose every day is a year, 365,000 years: and through all this mighty era those three matchless graces, holiness, happiness, and beauty triumphantly and universally reigning, and even the entire menagerie of Nature bathing itself in the mellow glory. Suppose, still further, that after he has sufficiently familiarized himself with the vision of this earthly elysium, and has just passed to and mastered the fact that, with a slight and relatively altogether insignificant break, this happy period shall everlastingly continue — suppose that another and still higher clairvoyance succeeds. His view is no longer confined to this earth. His eye has the freedom of the starry spaces. It sends glance outward and outward to find the voids peopled with worlds in such prodigious numbers and magnitudes that, in comparison, the great outspread of earth is but a point. What unspeakable legions, all cased in golden mail, go wheeling and charging and storming through the routed empires of Night and Nothingness! What infinite, infinite armadas, with flashing banners, bear down the reaches of that endless ocean — and behold all, with scarcely an exception, freighted to overflowing with beauty and goodness and bliss, as some gushing sunset cloud is freighted with the dolphin hues of the dying day! And he sees that the whole area flecked with sin and pain and various evils is comparatively but a fluxion of the last order, a microscopic dot on the white page of universal Nature. I say, suppose some second-sight could discover to him all this — should become the successful whipper-in of all its roving members to that august natural parliament in which the question of a good God is just now pending — bringing up substantially all space and all duration to add their voices to that large majority which on the earth utter affirmative suffrage — what would be the result? Would not the seven thunders of the ayes completely drown the noes in his ear? Ought they not?

Now, who is authorized to say that an actual canvassing of duration and space would not discover substantially all this? Not a man. Traditions favor a golden age to come as well as a golden age past. "Jam redit et virgo, redeunt Saturnia regna." The rapidly advancing sciences and arts and comforts of men point in the same direction. The magnificent faculties for virtue and happiness which every man consciously possesses — also actual examples, sometimes found, of individuals, families, and communities already well-nigh bright enough in every respect to enter into the composition of a paradise — look the same way with still greater steadiness and majesty. And then, what means the far superior aspect of most of those foreign worlds which sail so brightly and joyfully, and, many of them, with such marvelous glory, through the field of the telescope? Does that rainbow-bouquet of orbs in the Southern Cross, or that great cluster in Hercules which sails in such heavenly pomp across the field of our telescopes, positively discourage you and bid you think of abodes of sin and sorrow? Oh, no. They are a positive encouragement. They suggest a fairer state of things than we have here. They assert a possibility, they venture a prediction, they turn their faces hopefully toward the sun-rising; and, as we dimly look upon them, we imagine we see their features already beginning to light up with the flush of coming day. It is not from such facts that a Baconian infers discouragement. If he does it at all, it is from the evils seen in this world. But would a savage on the most barbarous South Sea island, after looking about his narrow home and observing what obtains there, be warranted in saying that, on the whole, probably all the rest of mankind are savages and cannibals, or that any of them are? Would a child living in the most dilapidated hut in Ireland, after looking about on its ruins and its rags, be warranted in saying that it is more likely than not that all the other dwellings of the world are as poor as his own, or that any of them are? Would a trilobite, after looking about his native marsh, be entitled to say that, more likely than not, nothing better than trilobites would ever appear in the world, or even that a single true trilobite would ever exist out of the Silurian?

If I have accomplished what I attempted, I have shown that the objection from the sins and sorrows and other shadows of the world does not lie against my thesis at all; that it is at a threefold remove from being pertinent even against the doctrine of a good God; that if it were intrinsically available for this purpose, it would still be balanced, heavily overborne, and, not improbably, completely sunk below the horizon by the actual state of facts in this beautiful and even gorgeous universe that surrounds us.

IV.

MACULÆ.

Πόσῳ μᾶλλον ὁ Πατὴρ ὁ ἐξ οὐρανοῦ.

Ως ταλλ' απαντα δευτερ' ἡγεῖται Πατὴρ Ζεύς.

Sophocles.

IV. MACULÆ.

FOURTH LECTURE.

MACULÆ.

IN the present state of religious thought, all objections to the goodness of God make directly against His existence. On this account I have taken pains to show that the maculæ of various kinds observable in Nature, are very far removed from being a valid objection to the Divine goodness.

This subject is so extremely important — the idea of possible malevolence in a Being of substantially infinite powers operates so powerfully to prejudice the mind against admitting His existence — that I propose to enlarge my answer still further. I propose to show that, despite all stumbling-blocks, the state of facts is such that, if we *assume* God to exist as the Author and Ruler of Nature, we are bound by Baconian science to admit not only His loving-kindness but a loving-kindness that is in the highest degree paternal. If He is at all, He is tenderness itself. If He is at all, never did sire so yearn over son as God yearns over all His creatures.

Let us, then, temporarily assume a God who is

the source of being to all other beings. Lo, the All-Father; lo, the Pater Mundi! More broadly and fundamentally than ever was man the father of a human child, is God the Father of all things, small and great, unintelligent and intelligent, lifeless and living, that people with their countless swarms the universal round of space — Father of the very primary elements, and basal substance of all things — Father of all natural chemical and mechanical combinations of these — Father of all natural structures; of the man; of the brute; of the plant; of the stone, whether as a jewel, a stratum, or a world Everything in Nature belongs to His family. Stars and souls are His children; the veriest insects and motes as well. You are His son, and so is the worm under your feet, as well as that atom of dust which the worm crawls over. There is not a thing which has not occasion to send heavenward its Pater Noster.

Let this be admitted. Then you are to observe that human beings are mere infants relative to this Heavenly Father. The greatest specimens of adult human nature ever seen; the men of broadest faculties, of widest information, of highest culture; the most famous scholars, statesmen, philosophers, geniuses — even such men as these are merest infants relatively to their Infinite Father. Compared with His faculties, what are those of a Newton or a Pascal! Compared with His knowledge, what is that of a Leibnitz or a Humboldt!

Compared with His accomplishments and feats of many names, what are those of admirable Crichtons and Sidneys and Cids! Mere nothings, surely. When I say that they are infantile, when I liken these so-called great men to the little children that creep and totter about our human homes, I certainly may be considered to speak with great moderation. We all know it an under-statement of the truth. So far from being hyperbole, it falls wonderfully short of expressing the actual facts.

Men are God's infants. And we ought not to be stumbled at finding them receiving from their Great Father what is found in our common household experience to be wise treatment for little children. I mean that such treatment as a wise human father finds necessary for or adapted to his very dear little children, it should not stumble us to find allotted by God to these very little children of His, adult men.

See how He treats us!

See, first, that *we do not have all our wishes granted.* How well do we know this! Why, it is only here and there one, among the multitude of our cravings, that God suffers to be gratified. Man is "a bundle of wishes," but he neither receives nor expects the fulfillment of the thousandth part of them. Let us confess it; had a chronicle been carefully kept of all the crude wishes that have flitted through our minds from day to day, we should not only be mortified at the quality of many of them and astonished at their number, but we should also be both mortified and astonished at the very small proportion of these blossoms which have ripened into fruit. — Well, it is but the case of the very little child in the hands of a wise and tender earthly father. Does he give his children everything they want — the little tottering, unreasoning, inexperienced, visionary things! He knows better than to do that. He has too much good sense and regard for his children to do that. He allows them to wish in vain for many a pernicious indulgence which he could easily give them if he thought best; even stoutly withholds such things from their tears and prayers. And when they have grown up they will be thankful to him for his wise and kind obstinacy. Is not God wise and kind after the same manner? Though we are men as compared with children, we are children, infant children, as compared with God. And not one in a thousand of our crude fancies as to what would be good for us is He disposed to fulfill. Perhaps He loves us too well. Perhaps He is too wise to do so foolish a thing, though our hearts cry bitterly unto Him for it.

See, second, that *we are positively stricken as well as denied.* Not only do we fail of having all that we wish — we also receive positive correction, chastisement, stripes. What man that lives is without his trials? What man that lives does not die — such is the hard word we use — driven out very painfully, perhaps, into the cold and dark? Losses, crosses — who has not looked many forms of such things in the face; nay, taken them firmly by the hand; nay, most reluctantly embraced them as men embraced the thorny Mater Dolorosa of the Inquisition? Is God therefore unpaternal? Is our case, after all, so very unlike that of other children? What son is he whom the judicious father chasteneth not? Does any wise parent neglect to act on that old-world injunction, "Correct thy son while there is hope, and let not thy soul spare for his crying?" Nay, the rod is not spared in any well-ordered household, in order that the child may not be spoiled. Sometimes, even his home is broken up, and he is sent out, sorely against his will, into what he considers the stormy cold and dark. He weeps, he wails, he suffers — suffers apparently as much as the man with his manly troubles. It is most touching, those distressful tones and features and contortions with which the little one shrinks back from what the parent decides must be done. "Poor child!" says the heart of the bystander. "Poor child!" say much more the softer hearts of sisters and mothers; and the moisture gathers fast in their eyes as they look on. It would be hard to show that yon yearling, drenched in tears and piteous exclamations, is not suffering as much as most dying men. Yet his father is firm. He carries through his plans as a business man, his plans as a household providence, his plans for training that particular child, without bating one jot. He transfers him from one school to another, from one physician to another, from one home to another; albeit it must be through a night of lowering looks, a sharp east wind of expostulations, and a free rain of tears. Is this treatment anything against the affection of the parent? Does any reasonable person conclude that firm father to be either cruel or injudicious? Perhaps every sensible, experienced man would think him cruel and injudicious if he should neglect that, for the present, painful discipline. — Now what are these grown-up men about us but merest children before God? And when we find the Heavenly Father correcting them after the manner of earthly fathers — a manner that we justify and even confess to be required by an enlightened and wise tenderness — why do we lift our eyebrows with complaining wonder? Is it any more than the usual treatment of well-loved and wisely managed little ones?

See, third, that *we have tasks and burdens put upon us which, doubtless, God could spare us, so far as mere power is concerned.* Cares, watchfulness, painful inquiries, various true work of body and mind, personal sacrifices of strength and time and property for the good of others — such things are imposed, sometimes very largely, on all our men and women by the present scheme of Divine Providence. — Well, is not this the way little children are accustomed to be treated by wise and tender fathers? Do not such fathers aim to accustom their children gradually to effort of body and mind — to think, plan, take care, conquer obstacles, bind themselves to diligence and order, task themselves at schools, ply various odds and ends of manual work about the house or the farm or the shop — true tasks and burdens, all of them, to childhood? These little burden-bearers are warmly loved. Pecuniarily, perhaps, their parents could afford to allow perpetual holiday. But they are too sensible and experienced and intelligently affectionate to do any such thing. Those children must have character. They must be prepared for a useful and honorable maturity. So they must bear the yoke in their youth. And those kind parents, without hesitation and with the high approval of all experienced lookers-on, proceed by degrees to impose that yoke according to the day and the strength of the little children. — Now, what are these grown-up people about us but so many merest children before God? And when we find their Heavenly Father laying upon them — laying upon us — tasks to do and burdens to bear which His almightiness could well spare us, in case it were good for us to be spared, shall we behave as though we have fallen on a very mysterious and stumbling state of things, a state of things that must be laboriously cleared up by besoms of both logic and faith before we can admit our God to be wise and kind? He, too, has the character of His children to look after. He, too, has their honorable and useful and happy future to provide for. Are we any better than little children in His presence? Why should He not give us the usual treatment of well-loved and wisely managed little ones?

See, fourth, that *we are always required to obey, often without reasons assigned.* Persons of ripe and even hoary years are not allowed to have their own way. The laws of the land say, No. Above all, the laws of God say, No. Bearing down most comprehensively on the lives and even the thoughts and feelings of the oldest and best developed among us, the laws of Nature, with their penalties, bring us the Divine wishes in unmistakable accent of command. Ye shall — ye shall not. No matter if we are kings, we must obey. No matter if we are sages, we must obey. No matter if we are venerable patriarchs, we must obey. Nor are reasons in full always given us for these commands. Sometimes there is only the simple expression of the sovereign Divine will. It is purely a case of unexplained and unexplainable authority. We cannot see why the law was established. So God has chosen; this is all we can say of the matter. — Well, in this respect we are treated like little children; as we are, before God, though our locks are silvered with age and wisdom. Are not wise and kind parents wont to insist on obedience from their little ones? Are they always careful to give intelligible reasons for their biddings? Obedience is the fundamental principle of all thrifty rising households. Reverence for parental authority, as such, is required. The narrow intelligence and experience of childhood cannot always have matters explained to them, but must learn to do things simply because the parent wills them. Do I bring certain strange things to your ears? On the contrary, are they not things that have been generally understood among thoughtful persons from the foundation of the human world? Do we blame these parents who insist on being obeyed? Do we pity these little children who must submit to authority? Not at all We blame the parents and pity the children if other principles are allowed. We know that both parties are in a fair way to ruin. And when God, our Heavenly Father, puts us who are called adults, but who are nothing more than little children before Him, upon a regimen of obedience, and strenuously insists upon it that, instead of doing as we please, we shall go by rules of His providing — sometimes unexplained rules — shall we wonder as if we had never heard of such things being done before by the kindest and wisest of parents? Shall we feel aggrieved and sore as to rights and liberties, as though we have not been heartily approving and commending, every day of our lives, just the same treatment of other little children by their earthly parents? What are we, grown up-men and women as we are — what are we but merest children before God?

See, fifth, that *we are kept in a state of close de-*

pendence on God, and under a necessity for daily appealing to Him for support, information, and guidance. You know how the Christian Scriptures put our case. It is God who really provides for us everything we have. He gives us our daily bread. He clothes us, as well as the grass of the field. Our education, our substance, our enjoyments, our honors ; in short every good and perfect gift, is from above, from the Father of lights. What have we that we did not receive? All things come of Thee, and Thou givest meat unto all : and unto Thee shall all flesh come! So we are to go to Him for everything we want — for the daily bread, the wisdom that we lack, guidance in the path we tread ; for, O Lord, it is not in man that walketh to direct his steps! Absolute and perpetual dependence on the Heavenly Father for everything, and a daily looking to Him for everything — this is the law of life to all of us, even the strongest and highest and proudest and most self-contained of our men and women. Now suppose this Bible account of our dependence to be the true account. What then? Is it a very stumbling matter, even to freedom-idolizing Americans? See how the little child hangs on his father's hand for everything! Everything is provided for him. Hector takes care in all directions; and whether the puny Astyanax is to be fed or clothed or instructed, it is the parental forethought and busy ministering hand that opportunely meet the needs of every passing day. The child has nothing that is strictly his own. For whatever he wants he has to go to another. So from morning to night he is saying in fatherly ears, "I am hungry; I am thirsty — what is it; may I have this or that; may I not do this or that?" In short, the father is the treasury to which the child looks and from which he draws, under such limitations as that father chooses to impose, every hour of the day. No property in stock is put into his hands from which to supply himself. From hour to hour he must appeal to the judgment and bounty of the sire. This is the law of our households, of the wisest and kindest of them. Is there any thing unreasonable in this, considering what little children are? Anything oppressive, harsh, unduly exacting, unnatural, considering what little children are? To be sure, there is not very much liberty, independence — as men sometimes use these words — in it; not very much of the principle expressed in such words as, "I do not care for you," "I am as good as anybody:" but there is fitness, order, safety, and a chance for happiness, usefulness, and religion in it. Who thinks the worse of a father for binding up his ignorant, inexperienced, incautious, and wayward child in such a system of daily dependence and appeal? You think the better of him for it. You would heartily condemn his lack of judgment, were he to take a different course. Is the man insane? Does he know anything whatever of the nature, tendencies, and interests of little children?

— Well, what are we, grown-up people, but little children Godward? And why is it not in the highest degree reasonable that our Heavenly Father should make our narrowness and inexperience hang daily and hourly on His wisdom and goodness for supplies, and should require us to go to Him with our asking for whatever we want? If this is a bondage, it is such a bondage as sensible men know is natural and necessary to the condition of little children. Little children cannot do without it. Their liberty has to be sacrificed to their safety.

See, sixth, that *we are not told of all the Divine affairs; that those we are told of are often allowed to seem inexplicable, unwise, and even unrighteous, especially to first glances.* Men sometimes complain because the Christian Scriptures are so silent on many points of curious and interesting inquiry. Much more show of reason have they to complain of the silence of Nature. Well, it is true that God does not see fit to answer all our questions, even all our theological questions. Some of His matters He keeps wholly to Himself. Others, of which we are allowed glimpses, are far from being well cleared up as to either the meaning, the wisdom, or even the righteousness of them. And many of His dealings and statements — for natural laws and providences are His statements — we are obliged to take altogether on trust. Does not the explanation, in part, lie in the fact that we are little children — our old men, our great men, our statesmen, our philosophers, and all — merest infants relative to the Heavenly Father? We are treated as all earthly fathers of average discretion are in the habit of treating their offspring during their tender years. Which of them tells himself and his affairs to the child of four or even twelve years, absolutely without reserve? Some things he keeps back because they cannot be understood, some because they would be misunderstood, some because they would be flagrantly hurtful to that early age. And such things as he does talk freely about — does he undertake the hopeless task of clearing up their every aspect to that as yet scanty intelligence? When it fails to see, as it often does, the full meaning of his conduct, or the good judgment of it, or the right of it, does he foolishly consume his time and strength on the impossible task of explaining and justifying his comprehensive and far reaching plans and movements to that glow-worm understanding? He knows better. However affectionate, he declines to do so foolish a thing. And may not God, though tenderness itself, decline to do the like? What are our maturest understandings in the presence of His great plans? What living man has breadth of view enough to take in anything more than the smallest angle of those Divine schemes and movements all of which embrace the universe and fill eternity? It is a matter of invincible necessity that sometimes Divine conduct, which really is fair and glorious as the day, should bear to us as mere gazers a very different aspect: it is only as believers that either the children manward or the children Godward can do full justice to their father, human or Divine. The man-father accordingly asks and expects his children to trust him where from the nature of the case they cannot judge of his conduct; and everybody says the demand is reasonable. And may not the God-Father also put His children on trusting Him in similar cases; and everybody be bound to say and feel that His demand is reasonable?

Such are sample maculæ. They fully represent the scope and weight of the whole class of natural shadows, umbræ and penumbræ, human and extra-human, for which God may be thought responsible. He is not to be thought responsible for the sad moral condition of mankind — as I shall, almost immediately, attempt to show. Assuming this for the moment, we have in those stern-featured ways of Divine Providence just cited the gist and essential variety of all those maculæ in Nature which seem to cast interrogation points toward Heaven. They are the gravest of all. In their scope they sweep the whole field of natural evil — at least this side of the essential constitution of man. If these do not mean anything as against even a paternal regard in God for all His creatures, there is nothing in the whole night-side of Nature that does. But they do not mean any such thing. See how much they are like the shadows of oui childhood. The kindest of fathers make these: why may not a Kindest of Fathers make those? Surely we ought not to lift our eyebrows in complaining wonder, when, being little children Godward, we find ourselves treated as little children by Him; treated as wise and loving earthly fathers are wont to treat their children with the general approval of mankind. Even the children themselves do not, in general, suspect either want of judgment or of knowledge or of love to themselves in such treatment. They may do it for a moment in a pet; but in general they possess that instinctive sense of their own narrowness as to faculty and experience which forbids their concluding against the father on such grounds. They trust and love him notwithstanding. They mutely say to themselves, "He knows best." They silently hearken to the filial instinct of trust within them which says, "He means it for good; it is the best that can be done under the circumstances." If a little child should be found habitually suspicious and sour toward his father on such grounds, every beholder would condemn the Phenomenon, and would not hesitate to pronounce him very unreasonable, very foolish, very unamiable, and very unnatural. It is felt at once that such conduct is the fault of a perverse, unfilial heart, rather than of a stumbled understanding. And why should not we condemn ourselves — we adult persons, and yet mere children before God — and say it is the fault of our wayward hearts, if we look with coldness and distrust on our Heavenly Father on account of such treatment as He gives us in common with all wisest and best of this world's fathers? We ought to know better. The consciousness of the mere nothingness of our powers, of our stark childhood and even infancy, as offsetted to the Divine plans and ways, should make these adverse seemings go for nothing. Shall we presume to be stumbled at the Heavenly Father for doing what is specially characteristic of the best class of earthly fathers, in proportion to their wise affection and solid greatness? Indeed, the maculæ are really faculæ; torches to illustrate the true paternal character of God. The harmonies, inductions, and Baconics of Nature interpret its shadows into lights.

But, thinks one, there is this great difference between the case of the earthly father and that of the Heavenly. The one has to accept and deal with human nature and its fundamental conditions as he finds them: the other had the making of this nature and its conditions. The human father is himself a creature, with very limited powers: the Heavenly Father is the Almighty Creator, to whose greatness nothing is impossible nor hard. That great Father could, with the greatest ease, have prevented the necessity for such unpleasant dealings by giving us a different nature, or by omnipotently manipulating that nature at the promptings of an infinite wisdom. Would any wise and kind earthly father subject his children to such unpleasant features of treatment unless he were compelled to do so? Is the Infinite Father compelled?

My friend, do you know what the word "Almighty" means? Do you not know that it means physical power? Compelled — yes, I reverently answer, compelled, in a sense and under the circumstances; compelled by His own wise and righteous heart. For, just consider. The nature which God has given man is the noblest style of nature known. It is even the noblest conceivable. It is a moral nature; capable of knowing, admiring, loving, freely choosing, and magnificently possessing and enjoying God and virtue in apparently ever-increasing degrees. No other nature is capable of so high an order of enjoyment as this. No other can glorify the Maker so much. The intelligent appreciation and voluntary homage of such a being must be the most precious and dear thing on which the Eternal Father looks down. What is the music of the spheres compared with that of a free, intelligent, loving soul! What are the glories of the day or of the night compared with the beauties and majesties of virtue! No, O Pyrrho, there is no kind of created nature so noble as that we possess. You cannot conceive of another as noble — with such glorious possibilities. Where is the man who is prepared to come forward and prove, I do not say to a demonstration, but to a probability, that God could have done a wiser and better thing than give us such a nature as this? Who knows it? Who will presume to say it? Indeed, are not the probabilities all the other way? — Well, if we are to have a moral nature, it must receive from the Creator a treatment in harmony with that kind of nature — must it not? It must have moral treatment; it cannot consistently be treated as a stone on a system of pure physical force. To what extent physical power can enter into the best system of moral treatment is evidently no easy problem. Where is the reasonable man who will pretend that the problem is easy, and is ready with his proof that probably physical omnipotence can enter that best system so largely as to make the case of the Heavenly Father with His children essentially unlike that of earthly fathers with their children? On the contrary, the cases must be essentially alike; because both contemplate moral beings under what is essentially moral treatment. Whatever minor differences exist, the treatment must in either case be, in the main, suited to the nature acted on. It must be genuinely moral.

So it appears that there is nothing in the Heavenly Father's way of dealing with His children but what should be allowed consistent with the largest measure of tenderness on His part toward them. Indeed, this way of dealing is, under the circumstances, a positive evidence of such tenderness.

This on the one side. On the other, behold, bedded in the constitution and course of Nature and as solid as any science that ever was studied, five Great Laws, which we have only to set in the light and breathe upon, in order to bring out on each in lustrous characters these words hidden in them from the beginning: SACRED TO THE FATHERLY LOVE OF GOD. LAUS DEO.

THE LAW OF THE INFINITE. We reverently approach the Divine Nature. We look at its knowledge, and lo, it is omniscience. We look at its power, and lo, it is almightiness. We look at its duration, and lo, it is eternity. If we proceed to look at its moral character, shall we not continue this finding of immensities — shall we not find its moral traits laid out on the same grand scale as its natural? It is to be presumed. Proportionateness and equilibrium are the habit of Nature.

The Divine Nature is eternal. This means an eternity of moral practice — good or bad. And this, according to the way of all the moral natures we happen to know, means in God a present goodness or badness as colossal as that Past through which it has been exercising. It has acquired an infinite momentum in moving down that long and mighty slope. It has the habit, the inveteracy, the solidarity, the amplitude of innumerable chronologies. And, to-day, by virtue of that amazing practice, God stands at the head of the universe, either the best Being in it or the worst. He burns toward His creatures with an awful malevolence or with a benevolence more glorious than the brightest of His myriad suns.

His infinite reason and conscience imply the same thing. He sees with unbounded clearness the unbounded beauty and majesty of virtue, and its unbounded importance in a Being clothed with such powers and occupying such a position as Himself. It is in the very focus of His omniscience that He is under infinite obligation to gloriously love His creatures and do for them according to His splendid opportunities. Their very helplessness in His hands is itself a piteous appeal for gentleness and tenderness. He meets that appeal and fulfills that obligation, or He does not. If He does not — if in this blaze of manifestation He neglects to play His magnificent rôle in all its entirety, notwithstanding it involves not the slightest difficulty to Himself — He incurs an infinite guilt. To persevere in this course, on through the measureless stretches of Forever, despite the beseechings of such an intelligence and of a universe whose every need is always before Him — what a stress and audacity of depravity it requires! From the nature of the case it must be stupendous; from the nature of the case it must be incalculable. Such another sinner the universe does not hold. He deliberately sacrifices, age after age, in the face of infinite light, an infinite good in virtue itself and in all the glorious results which perfect virtue, armed with omnipotence and omniscience, could secure. So He is infinitely righteous or infinitely unrighteous, infinitely benevolent or infinitely malevolent.

Which is it? This moral character which, whatever it may be, harmonizes in its proportions with the other parts of His nature — is it that glorious goodness which insures toward all His creatures a grand paternal tenderness, or is it that appalling badness which insures to them the government of an infinite demon? Look about you. Is it such a government you and I are living under? Is the state of this world as bad as almighty malevolence could make it? Suppose a fiend, panoplied in omnipotence and omnipotent subtlety, to set himself to make the universe as corrupt and miserable as he could; would he get no nearer his object than such majestic heavens as those, and such fair and on the whole, happy world as this? Preposterous. So we must take the golden horn of the dilemma. God is as much the best as He is the greatest. The Father is the best of fathers. He is tenderness itself toward His children: and all His shaded measures with us are as truly conceived in an exquisitely benignant spirit as are those other measures whose radiant faces pour delight on all eyes.

The Law of Conscience. One part of the law of conscience is that pain shall follow conscious wrong-doing, and pleasure conscious right doing. Another part is that the father who hates or is indifferent to his little children shall be deemed cruel and monstrous. The pressure of this law is universally felt. The Maker has evidently framed t into the constitution of mankind. Make your inductions; you shall, in the regular way, find it as much a law of Nature as is the law of gravitation. You can overcome gravitation and go away aeronautically from the earth, instead of going toward it. So you can overcome remorse for known sin and make it a remorse for holiness; so you can overcome your horror of the man who hates and tortures his own little children and perhaps convert that horror into a liking. But, for all that, the horror is as natural to man as gravity is to matter. So is the connection of pain with conscious wrongdoing. Both are cosmopolitan. Both belong to human nature. No clime nor country nor class nor culture nor capacity nor condition where they are not at home. How came this? Did God give these things to human nature because He could not help it; because He could not have a human nature without it? Nay; there are just enough triumphs over the law to show well the possibility of dispensing with it. Men do sometimes succeed in reversing the poles of conscience, and come to feel only pleasure in vice and pain in courses that look toward virtue. What individual men do for themselves sometimes, God might have done for the race always. He might have set the needle of the entire human conscience astray from the beginning. Had He so chosen, He might have made the universal moral sense of the world sing jubiees over sin and dirges over holiness. Why did He not? Doubtless because He wanted to have men virtuous. Natural laws tell God's wishes as plainly as any speech can do. When we are benevolent, He says sweetly in our ear, "Well done"—more sweetly than Orpheus or Apollo ever sang. When we are malevolent, He says bitterly in our ear, "Ill done"—more bitterly than ever rue or wormwood spake to the tongue. It is because He greatly wants men the world over to love rather than hate. This shows what He is. Of course His own character is congruous with His wishes. He who profoundly wishes all men to love each other, Himself profoundly loves all men. He has Himself the virtue which He so earnestly, persistently, and universally wishes others to possess. He loves it, after the fashion of a Divine Nature. He lays Himself out for it and its affiliated felicities, in all this human domain, after the fashion of a Divine life. He is a true Father. His good-will to men is splendidly paternal.

THE LAW OF PATERNITY. Look about you among human parents. Do they not, almost without exception, tenderly love their little, dependent, helpless children? Is not that person considered almost a monster who approaches a state of heartlessness toward those who lisp toward him the name, father? Is not parental love an instinct through all the graded parentage of the brute kingdoms? The birds, the quadrupeds, the fishes—animals, domestic or wild—how the feeblest and most timid of them will flame forth in reckless self exposure to defend their young! Even the philosopher who is parent of an ingenious theory, the author who is parent of a creditable book, the inventor who is parent of a useful machine, the discoverer who is parent of an important science or fraction of a science, the artist who is parent of an excellent statue or painting, the statesman who is parent of a wise measure of national policy, the mechanic who is parent of a beautiful ship or house or watch — all such persons find themselves having an affection for the things toward which they sustain this relation of paternity. They are the root from which that beautiful greenness has come: their image is on it; their life is in it; their body, their soul, their genius, their patience, their knowledge, their character, is diffused through it; in short, they have in all those green leaves and yellow fruits so many promising little children of their own. And they almost invariably conceive an affection for them as such. The abuse of them is the abuse of themselves; the praise and beauty of them are the praise and beauty of themselves. Such is the law of paternity everywhere within the range of our observation — the parent loving its offspring. Among all the animal tribes, in earth and air and sea — whether the child be flesh and blood; or only the cell of the bee, the nest of the bird, the dam of the beaver, or the hand-work, mind-work of the ingenious man — it is loved by ts author. And now we have to ask, Is God the sole exception to this sweeping law of paternity? Is the Being who established this law, and armed it with flails of remorse — is He Himself out of harmony with it? Does He, too, not love His children, whether their name be men or oxen or birds or flowers or oceans or stars? The induction, the science, is overwhelmingly against it. It is, in fact, against much more than this; against that parental regard in God being anything short of a most exalted and permanent principle. For, looking around, you observe that in all cases such a principle lasts as long as there is occasion for it; as long as the care of the parent can really be of service to the child. How long, pray, can the care of eternal and almighty God be of service to man and the other creatures! Looking around, you observe that the higher the grade of the parent, the more elevated and enduring his attachment to his offspring. In man it shows its noblest quality, and in man it lasts indefinitely. Surely, in the supreme God we should look to see the principle shine divinely and shine eternally. Looking around, we observe that the higher the grade of the offspring, and the more completely it springs from and depends on the parent, the greater the affection which that parent expends upon it. The more valuable the discovery which a man has made, or the machine which he has invented, and the less he is indebted to other sources than himself in the production of it, the stronger his regard for it

Well, we should accordingly expect that man, who is chief of God's works and children in this world, so far as they are visible, and whose whole nature, body and soul, substance and organization of substance, had its origin solely in Him, and hangs totally on His hand — we should expect that man would be favored above all the other visible children of God with His fatherly regard. It is, in fact, but another example of the use of that famous Baconian induction which has built up our modern sciences. Are these sciences good for anything?

THE LAW OF CHARITY. This law is, Assume a person innocent till he is proved guilty. For example, assume a man honest till you have positive evidence that he is dishonest; assume that a man is not a murderer till you have positive ground for believing that he is a murderer; assume that a father is paternal in his feelings, till some positive reason is found for believing him unpaternal.

It would be monstrous to go on the principle of treating a man as guilty till he is proved innocent; to treat him as possessed of all the vices till he has proved himself possessed of all the virtues. It would be bare justice to withhold positive condemnation, and treatment to match, from the man till he is proved guilty. It would be charity to consider and treat him as innocent till he is proved otherwise. On the one hand, till such proof is brought, justice requires us not to decide against him; on the other hand, till such proof is brought, charity requires us to decide for him. Where conduct is equally well explained by two hypotheses, we are to take the most charitable one, instead of taking the least charitable, or instead of holding ourselves neutral between the two. We are to do as the spirit of kindness would prompt.

Such is the law of charity — a law that has forced itself into recognition and supremacy in all decent judicial proceedings the world over; a law on which the theory of social intercourse has come to found itself without contradiction in all well-ordered and enlightened countries; a law which the humblest among us knows of, and, on occasion, claims the benefit of as a matter of commonest right rather than of charity; a law which is the natural prompting of a kind and friendly heart; a law born of the Golden Rule, Do to others as ye would that others should do to you; a law found in practice most convenient, safe, fruitful, necessary — saving a world of vexation, mischief, and cruelty — in fine, a law which, while not against justice, is sublimely beyond it; something gloriously higher and completer; in fact, righteousness.

Now this great law requires us to postulate the paternal love of God. We are not to withhold from Him the benefit of that generous principle which we concede, at least in theory, to all our fellow-men. We are to give Him credit for being paternal in His feelings till He is proved unpaternal. Can He be proved so? The attempt has been made; but we have seen that the most unpleasant features of His dealing with us, so far from being of the nature of an attack on His character as a Father, are not only perfectly consistent with but even suggestive of a wise tenderness on His part. Under these circumstances, the law of charity steps in and demands of us that we put a favorable construction on appearances; that we take a friendly and generous view of the case; that instead of judging our Maker and Father from the side of harshness, or from the side of indifference, we judge Him from the side of good-will; that we say to ourselves, "He shall be esteemed innocent till He is proved guilty." "He shall not be suspected even, till there is made out positive ground for suspicion." "We will do to Him as we would that others should do to us." If this is more than just, it is not more than righteous.

THE LAW OF THE GENERAL RULE. A child is in such close and constant dealing with his earthly father that he cannot well hold himself in a state of suspended judgment as to whether his father loves him. He must decide. Well, if he must decide, a very natural consideration to present itself is, Does he treat me as if he loves me? And, in case some facts apparently look one way and some the other, it is very plain that he ought to decide the case, not according to the exceptions of treatment, but according to the general rule. It would be plain absurdity, if a decision must be come to, not to base it on the general tenor of the treatment received. If this is found to be as if his father loved him — if he finds that, in general, his father treats him as though he were a dear son — he ought to admit that such is really the fact, though he cannot explain in consistency with it a few facts that have a contrary look. He is to go by the rule, not by the exceptions. Suppose, then, he finds this to be the state of facts. He hears his father say that he loves him — not unfrequently. Not unfrequently he finds the parent directing towards him kindly and tender looks, smiling upon him, and even embracing him. He looks around and finds himself sheltered in a beautiful home, and sees specially assigned to him his own beautiful rooms filled with conveniences and beauties. He finds himself wisely and abundantly fed and clothed and instructed by his father; especially finds himself taught by him carefully and well on moral and religious matters, including the reciprocal duties of parents and children; finds that his father tries to keep him from all evil ways, and to give him those principles and habits which are fitted to secure him a happy, useful, and prosperous manhood; finds that he so deals with him that he is actually, on the whole, happy, and would be vastly more happy if he conformed as carefully as he might to all his father's laws and hints — indeed in such a case would surely become a most happy, useful, and noble man ; in addition, as I have already partly said, finds him from time to time calling him by every manner of endearing name and epithet, nursing his sickness, comforting his sadness, saying that he loves him, assuring him that on occasion he could and would freely die for him, promising him that if he will try to do well the abundant ancestral riches shall provide more magnificently for his mature life than he can now possibly imagine ; finds that such things express the general drift of his experience as a child. Ought he to be at a loss how to decide his problem, though he is at a loss how to explain occasional severities of dealing on the part of his father? But if he is at no such loss — if he plainly sees that these things of exceptional aspect are or may be, after all, but the natural expression either of an invincible necessity or of a wise tenderness in the father — much more readily should he accept a conviction of that tenderness. Under such circumstances the child always does accept it; indeed, always does so with merely a vague notion and instinct of these facts. Does he not do rightly? Would he not be universally condemned were he to do otherwise?

To apply the illustration. Situated as we are, we cannot avoid taking up some positive attitude as to the question, What is the feeling toward us of God our Heavenly Father? We must make a judgment. And, in order to do it, a natural question is, How does He treat us, in the main — is the general drift of His dealing with us kind and tender? Allowing that, here and there, unexplainable facts of adverse appearance exist, it were monstrous to decide according to these in defiance of the mighty majority. Better to defy the pitiful minority. Dark-browed and resolute as these exceptions may seem, if we go against anything, let us go against these in their weakness and scantiness. These are a few stragglers; the others a compact and disciplined host. For, look at them! First, the Great Father professes to love us. To lay no stress on hundreds of written declarations, "I love you, I love you," professing to come from Him; the flowers, the songs, the golden light, the precious perfumes, the delicious tastes, all grace and beauty of form and feature and motion abroad in Nature — these are so many loving words, smiles, caresses of the All-Maker and All-Father toward the intelligent creatures who are aware of them. Who does not know it — vocal speech is not the only language, the lighting up and wreathing of a face is not the only smile, the pressure of an arm of flesh is not the only embracing! O bright-faced sky, O smiling earth, O scented and singing and festival springs and summers, O innumerable anthems and poetries of delightful Nature — ye also are God's tender looks and words and sacred kisses to us! Next, see what a beautiful home He has fitted up for us; ceiled with sapphire, floored with emerald, walled and curtained with sunsets and sunrises, upholstered and garnished superbly and almost unboundedly for our shelter, our convenience, our dignity, and our delight. Then see what stores of heathful and pleasant food he provides for us in the manifold grains and fruits; of suitable and comely clothing in the bolls of cotton, the fleeces of flocks, the moils and spoils of the silk-worm; of useful and exalting knowledge in the eyes and ears and other organs by which He puts us in communication with the wonders of this wonderful universe. Especially, see how careful He has been to furnish every man with a conscience to inform him and reform him on matters of right and wrong; and even to assure him that the Creator ought to love, and would be greatly guilty were He not to love, and, to the extent of His power, bless His creatures. See how, by means of conscience, and the laws of Nature, and the general strain of providence — to lay no stress on the written laws with their impressive sanctions which claim to have come from Him — He seeks to influence us virtue-ward and shape us to good principles and habits; for, let it not be forgotten, all intelligent observers admit that the general flow and pressure of nature and experience are in favor of goodness. See you that most men are actually secured by His care so much happiness that they had greatly rather continue to be than cease to be, though annihilation were made painless or even pleasurable. See you, also, that it is consciously in the power of every man to be vastly more happy and useful and noble than he is, by carefully improving the opportunities furnished him and carefully conforming to such Divine laws as he can discover ; indeed, in his power to be gloriously happy and noble. See the Great Father dowering us with imagination and hope that kindle and expand in a course of well-doing, and — while conscience says, "Dear one," and sacredly fondles you in God's name ; and the Scriptures say, "He has loved you enough to die for you" — see how that wonderful fancy and hope, conscience-prompted, begin to prophesy in vaguely magnificent speech of unspeakable glories that flush and span with their triumphant arches your ascending way. See such things making up the general rule of treatment from our Heavenly Father. Oh, if we could say nothing in explanation of the occasional somberness of the paternal Heaven that bends over us — if it were thickly beclouded beyond all falcon glances of our wisest and best — still our judgment should refuse to be ruled by the poor and scanty exceptions, and should bow instead to that kingly rule whose crowded congregation of voices proclaim in sonorous harmony a loving Father in heaven! But since these exceptions are explainable and explained — since it appears that they only show a dealing common to and characteristic of a wise and tender human fatherhood over moral beings — we yield ourselves with redoubled cordiality to the law of the general rule, and say fervently, God our Father does love His little children.

Altogether and law upon law, I discern an Unlimited Love and Righteousness outbeaming from the heavens. An unlimited righteousness! This attribute completes in God a perfect and glorious competency to govern. Wisdom, power, and duration without measure — surmount this cathedral structure with the superb roofing and dome of a perfect goodness, and you have a wondrous palace from whose golden gates may fitly issue the edicts of a universal monarchy. Behold a God who is abundantly able to manage the affairs of a universe, and fill its august throne to infinite advantage! That a system of things made up of blind matter and fallible intelligences and depraved hearts should go on as well, or a millionth part as well, by itself, as under the scepter of such a complete Being, is incredible and impossible. So He ought to govern. So it would be an unspeakable wrong to His creatures should He neglect to govern. So, gloriously perfect Being as He is, He surely occupies the supreme throne over all, as His magnificent duty and their magnificent necessity; giving to the different sorts of things in the universe the kind of government suited to each; giving to blind matter the government of physical omnipotence, and to moral beings, with their mixed constitution, a mixed government of physical force and moral law. In which case the actual state of the world, with its lights and shadows, its smiles and its tears as well, is throughout the expression of a grand Divine love.

This on the *supposition* that God exists and is the Author of Nature. The scientific corollary of this supposition we find to be a fatherly love of Divine proportions beaming on the world. And the maculæ of all sorts, properly attributable to God, so far from making against His goodness, are themselves parts of a broad and consummate scheme of love by which the Heavenly Father ministers to His great family.

V.

IN TENEBRIS.

Αριστερὰ καὶ οὐ κατέσχον, δεξιὰ καὶ οὐκ ὄψομαι.

Ποίησον δ' αἴθρην, δὸς δ' ὀφθαλμοῖσιν ἰδεσθαι.—*Homer*

V. In Tenebris.

FIFTH LECTURE.

IN TENEBRIS.

ANOTHER objection to the existence of God is drawn from His OBSCURITY.

It is claimed that a God would have made Himself more apparent to men than He seems to be. He would have so manifested Himself as to make it impossible for men to doubt or to neglect His existence. At least He would have done more in this direction than is observed. A large class of men find small difficulty in living with little or no veneration for such a Being, or even thought of Him; the faith of many in His existence is troublesomely weak; some have no faith whatever; while others positively disbelieve.

In gradual answer to this objection, I submit the following considerations: —

1. *A perfect revelation of God to human intelligence would be impossible.*

Of course, a finite being cannot fully understand one who, relatively to himself, is infinite. Were God to undertake the task of revealing Himself to me in all His completeness, in order to succeed, He would be obliged by the nature of things to expand my intelligence into dimensions like His own. Even as only a philosopher can well understand a philosopher, so only a God can fully understand God. There is no help for it — to beings so narrow as men, God must be, in large part, shrouded in impenetrable obscurity.

This fact, of necessity, precludes men from the most impressive view of God; namely, that of His whole nature, with its entire wealth of resources — a view, of course, infinitely more magnificent and memorable than that fractional view that is possible to us. It also, of necessity, precludes men from the best class of natural evidences and illustrations of the Divine existence; that is, those broader and more complex plans and works on which He has laid out the most wisdom and power. A perfect Framer of the universe must have one plan that is all-embracing; in which each thing is so delicately framed into every other thing throughout space and duration as to make of the whole mighty complexity a glorious unit. Who but a God could master such a scheme as this? The problem of the two bodies mathematicans have solved; that of three bodies yet bids defiance to analysis; that of the solar system, much more that of our galaxy, much more still that of the stratum of nebulæ in Virgo, is one which we never expect to solve. But how much further beyond us still is that universe-system, in which every atom is to be considered a world, and every world is to be considered, not merely in its mechanics, but in its metaphysics and its morals! But if we only could grasp this all-comprehending system in all its detail of harmonies and wisdoms, we should, no doubt, have such a regal evidence of the existence of a Divine Mind as would dwarf into invisibility any natural proof of Him which may now be within our reach. Our narrowness cuts us off completely from this grandest proof. It can even make that proof seem in parts positively ill proportioned, incongruous, empty, and even noxious. We are like the little child of a statesman. Some of that father's smaller domestic arrangements he understands — perceives them to be right and wise. But those great national and international schemes and movements are all nebulous and confused to the unfledged thinker. They look inexplicable. He sees no wisdom in them. He sees here and there things of apparently an opposite character. And yet this is the man with the fame of whose statesmanship the country is ringing; and these the very works of his which shall entitle him to his historic place among the astutest managers of empires. The child, because he is a child, is compelled to miss what is really the best illustration of his father's greatness.

Thus the very finiteness of our faculties suspends a veil between us and God.

2. *A revelation of God much short of what our intelligence could grasp would consume us.*

Suppose God should take as much of His nature as we can understand, and bring it to us in the way of adequate personal manifestation to the senses. Could we endure the exhibition? Why, we can hardly bear such sights and sounds as we ourselves can produce. We ourselves can kindle such glory of conflagration, can detonate such might and majesty of sound, as shall destroy sight and hearing, and even shock the weakly out of life. And surely if God should come upon our senses with such imperialism of sights and sounds as would appropriately represent the utmost power and knowledge we can conceive — as would worthily express our ideas of Eternity and Almightiness and Omniscience — that moment would be our last. Appropriate to the utmost power we can conceive — why, we can conceive of a power that can take up the isles as a very little thing, poise a planet in its hand, launch forth a nebula on the void as if it were an atom, crush together instantly into one palpitating pulp all known sidereal systems with as much ease as you do the beaded cobweb of a summer's morning! We could, if need were, imagine a power equal to still greater feats than this; one, with reference to our powers, properly called infinite. And I say what a show it would have to be to worthily express, not merely such a power as that, but also a virtual Omniscience and an absolute Eternity! Could our senses or even our lives endure it? Even now, when the common lightning shoots before our eyes, how they quiver back from the blinding flame; and when the common thunder comes upon us in some great crash, how our ears and hearts quail under the terrible bass! And were God Himself to come flashing and pealing on the world in all that outward majesty that rightfully belongs to Him, and fitly signifies to sense the presence of a virtually Infinite Being, who of us would see another moment in the body? We should straightway be dazzled out of life. Our frighted senses and hearts would give one leap, and then become motionless forever — and this though they were a thousand-fold stronger than they are. The men who ask that God should personally manifest Himself to their senses as God, know not what they ask. Do they want to be driven out of the body by the fire and sword of intolerable discoveries? — After our finiteness has hung one veil between us and God, our safety as embodied beings would compel Him to interpose another.

3. *A revelation of God such as would not consume us, would yet so shake and derange the mental faculties as to prevent due use of the revelation.*

Scenes not sufficiently rousing and awful to shock men out of life, are often enough to shock them out of reason. How many have been made idiotic or insane by appearances which took fierce hold of their imaginations and astonishments, and by them so shook the soul that it fell into dreary ruin! That terrible fire; that fearful explosion that awful storm or battle; even that strange flight of meteors; that portentous comet; that bloody sun; that crude marvel of the Spiritualist or expected marvel of the Millerite; that apparition of some sort, from the airy colossus of the Brocken to the ghostly form in the weird moonlight — how it shattered the man! His body survived, his senses escaped without harm, but the more delicate system of nerve and brain and rational thought could not endure the strain, and became a melancholy wreck. And so we see that it is not enough for God to abstain from such manifestations of Himself as would destroy the body. He must also abstain from that lower manifestation that would break down the stamina of the soul, confound the power of rational judgment, and paralyze our faculty for using a manifestation. Should a being of virtually infinite glory and majesty come on our senses, or our thoughts, with any but the most inconsiderable fraction of His greatness, our feeble souls would infallibly go into unhingement. I am now speaking of the most compact and stoutly built souls. I would not have trusted a single one of them, though bearing a supreme name among hero-worshippers, in the midst of even such phenomena as belonged to the breaking up of the Paleozoic or Mesozoic Period. Such outpour from above; such heaving from below; such rendings of the caves of Eolus and of Neptune; such tourneys and concussions of the enfranchised winds and waves; such thunderous

Marseillaise of the wrathful mob of volcanoes, earthquakes, tornadoes, and oceans; such an awful mäel strom of blackness, blaze, sound, and death; such wars of the Titans with the Gods — must have twice exterminated every overt species of hearing and seeing animal life; first by affright, and then by direct violence. Had man been there, though his soul had been boned and thewed like Alcides, he would have lost his reason. These delicate nerves and brains of ours, that so bow and break at the approach of what, after all, is mere Nature — what would they not do at the approach of the Great Supernatural in any fitting circumstance! But care must be taken not only for these few heroic souls, but for the much larger number who are no heroes — for the many frail; the many sick; the many excessively timid, nervous, superstitious; the many ignorant, weak-minded, ill-balanced persons scattered everywhere through the world. For the sake of these — for the sake of those hundred neighbors of yours whose souls are greatly less firm in texture — God must still further limit the public manifestation of Himself to the senses. He does not want to turn the world into a Mad-house any more than into a Morgue. There are already enough lunacies and unsoundnesses of mind in human society. — So another veil must be interposed between man and God. The man who asks that there should be made to us a full discovery of God, knows not what he asks. Does he want the world peopled with Gods instead of men? Does he want to become a handful of ashes, or at least to be dazzled intc corpsehood, by an insufferable brightness? Does he even want his reason and nervous system to fall to pieces under a manifestation of the Eternal? If not, he must be content to have at least three veils hang between him and God.

4. *Such a revelation of God as would not derange our minds, would speedily benumb the faculty of astonishment and general sensibility, and so would soon cease to be specially impressive, and could only be resorted to occasionally.*

If Deity is local — if His personality is not universally diffused, but occupies a limited district, such as Christians call heaven — then He could not make a permanent personal manifestation of Himself here without permanently depriving other parts of the universe of a manifestation. The most He could do would be to supply some standing substitute for Himself; perchance some astonishing Form to lighten through the sky and personate that Divinity which for the greater part of the time it does not include. A sincere being cannot be supposed to do this. The best He could do would be to make an occasional personal appearance among us; and for the rest of the time supply such august messengers and other miracles as should testify of God to human senses without purporting to be God, or to include His actual presence. This is really the best that could be done. But, if you please, we will suppose it is not — we will suppose that both of these modes of revelation are permanently open to Him: every day a mighty Form that really includes God can send its dazzling pageant on our sight, and every day the angels can fly and the dead rise and Nature tremble in awful testimonial to His being and greatness. How would such a system as this work?

As we have seen, the grandeur of the exhibition must be greatly moderated, to make it safe for either the senses or the intellects of men. But, within the limits of safety, a very grand exhibition might, doubtless, be made. We can be greatly moved and astonished without any danger to reason. But we cannot remain greatly astonished for any length of time, especially by the same thing, not even by any variety of things. As men now are, a permanent astonishment is impossible. No wonder can continue a wonder save for a very brief period: and the greater the first effect the sooner it will be over. Can Niagara astonish and awe you indefinitely long? Ask those who have lived out years by the side of it. Try it yourself. As soon as you perceive the effect decaying, pass to the ocean, and you may in a degree renew your feeling; but how many months or even weeks will it be before the sonorous majesty of the Atlantic itself will abate on the familiarized senses! Then take the traveler's privilege, and go on as swiftly as steam can carry you to another wonder, and then another. By a timely passing from the cataract to the main, from the main to the Alps, from the Alps to Latin museum and palace and cathedral, plethoric with the glories of genius and antiquity; from these to the astonishment of the Pyramids and Carnac and Luxor, you may protract the interest very considerably; but at last, and that after no long time, the faculty of wonder will become so numb that you can witness the most remarkable object with as little movement of soul as any commonplace object is wont to inspire. You will turn your face homeward. Why should you go on when your heart has become a mere clod — when you would hardly turn a corner to see Thermopylæ, or climb a hill to find Olympus in session! Such is the common history of sight-seeing. And I make no doubt that, were God to appeal to the senses of men daily by astonishing revelations, it would not be long before we would be as little impressed by them as we are now by the daily sun or nightly dome of stars; as the men were who once actually supposed the sun to be God, and in many a Heliopolis thought every thunder to be his voice, every whirlwind his breath, and every earthquake his movement; as we theists are now by the constant advent of bodies and souls into the world without any apparent cause that does not seem to us infinitely inadequate. Variety in the form of revelation would protract the first astonishment and awe; but despite everything they would speedily come to an end. Those born and bred to such displays — and such should we all have been had the principle of a permanent display been acted on — would never have any special impression at all from the daily grandeur, any more than those who have always lived at Chamouni or Eddystone, in daily view of the highest majesty of the Alps or the highest majesty of ocean. In order that any safe astonishing display in behalf of God may have the maximum of effect, it must be only occasional.

As to just how frequent the exhibition could be with the best advantage, it would be hard to say. No theory of maxima and minima with which I am acquainted solves the problem. But I am inclined to think that the man who made the round of the seven ancient wonders of the world once in twenty years, got more impression from them than he would have done by seeing them once in five. And I know that the friend who made one voyage on the ocean and saw it in all its moods, and then was ever after left to his memory and imagination, carried with him to his grave a grander sense of the huge flood than his companion of equal native sensibility to such things, whose whole after-life was spent upon it; or than that other equal companion who made his second and third voyage. And what right have I, as a logician, to venture on the affirmation that a single astonishing exhibition in favor of God would not do as much with most men toward placing Him in an impressive light as any larger number of such exhibitions? Indeed, as facts stand, I would not like to affirm that with most a mere tradition of such an exhibition, well told and well believed, and then committed to that wondrous painter, the Imagination, would not, on the whole, be more impressive and just and durably influential than sight itself. Never such an Apelles as the Imagination. She has colors on her palette, and models in her eye, such as never enter into pictures on the retina. She habitually outpaints all the galleries. And she can, out of her own resources, give a brighter picture of God and miracles than could possibly, with any safety, burn its way through the lenses of the eye and the labyrinths of the ear. And let this picture have the prestige of reality — let it be fully believed in as expressing substantial fact — and it will surpass all other pictures in impressiveness as well as brightness. It will also surpass all others in general truthfulness — provided its outline is really fact. No doubt sight is the truest painter of common objects; but not of such as have immeasurable greatness and excellence. We cannot get too impressive a conception of these. The nobler the conception, the truer. Let that supreme colorist, the Imagination, lay out all her power and dip her pencil in the sun; it will be an incomparable picture of God and astonishing miracle that she will give, and as much juster than all others as it surpasses all others as a mere picture. And, further, a picture by the Imagination does not lose its impressiveness by repetition, as do the pictures of sight. One hundred sights of Niagara will practically abolish the wonder; a hundred sights of a miracle would practically make it no miracle; but not so a hundred imaginations of these things. For, unlike the senses and the nerves, the Fancy gets more skill and power with every picture she makes.

Now, this greatest of painters, whose canvas is so impressive, so true, and so durably efficient, can only work to advantage under certain conditions. A flood of light on an object, of course, takes it completely out of her hands. Sight, in any of its degrees, restricts the freedom of her pencil. It is only when an object is given in mere outline to faith — and sight never gives an object in mere outline — that she has the fullest scope and motive for all the wealth and witchery of her art. She then has the inspiration of faith stimulating the highest freedom of invention. No, indeed — I would not venture to say that the credited tradition of a Divine manifestation would not, with most persons, be better than the sight of it; that, for most, to read of Divine manifestations and miracles in perhaps distant times and countries, and believe in them, and then to leave them in the wonder-working hands of that faculty that has built up the world's great epics, would not fasten on the mind the justest as well as the grandest sense of them, and warrant the saying, "Blessed are they who have not seen, and yet have believed!"—Evidently, we must consent to have another veil hang between us and God. He neither wants to turn the world into an Olympus, nor into a Necropolis, nor into a Bedlam, nor into a Bœotia. Shall He make men stare till they are stupid — make wonders so familiar to them that they must cease to be wonders? Whoever asks for an unstinted revelation of God, knows not what he asks. Does he insist on being himself a Divinity? Does he want his senses and his life to shrivel up before an intolerable glory? Does he want his frighted soul to leap into the abysses of distraction or idiocy? Does he even want prodigies wasted by a vulgar frequency; or by acting on the principle that, for most, the great spiritual powers, Faith and Fancy, with their sunset pencil and immeasurable canvas, have less power to render the Infinite to the soul than has the gross bodily sight with at least three dense curtains before it? If not, he must be content to have four veils hang between him and God.

5. *Such a revelation of God as could be permanent without benumbing our sensibility, must be still further limited by our depravity.*

If men were perfect in point of moral condition, the preceding causes of obscurity would still act. But men are not morally perfect; they are very far from it; their moral natures are universally and wretchedly broken down and corrupted. And this fact cannot do otherwise than heavily becloud the Supreme Being. Who expects a sick body to do bodily work as well as a sound one can do? Who expects a muddy pool to reflect the sky as well as the fountain of Helicon? Who expects the astronomer gazing at a heavenly body through a cracked and unhomogeneous lens to see it as well as through a Clarke or a Frauenhofer? It is not possible. Nor is it possible for a sick, turbid, broken soul to give as clear and just views of God as would be natural to an unfallen being. Sinners can see trees as well as if they were no sinners: doubtless can see outward prodigies expressing God just as well as they could with perfectly pure hearts. But as to seeing God in these things — this is quite another matter; and it is just here where sin acts as an obstruction. Sin is naturally averse to and afraid of a perfect Deity, naturally unwilling to recognize Him, naturally glad to find pretexts against the validity of any evidence of Himself which He may furnish. It is disposed to shut the eye and ear on such evidences, to look for them anywhere but in the right place, to see what it must see with the dead eyes of statues or with the dwarfing eyes of insects. In the presence of such a disposition no Theistic illustration, nor argument, nor even ocular demonstration, can pass for what it is worth. In the presence of the higher degrees of such a disposition the brightest Divine credential — though its name be Alcyone, central sun of the whole system of evidence — must be sadly blurred, and may be totally suppressed; indeed, is not unlikely to be totally suppressed. Why, the man has only to say, "Beëlzebub," "magic," "optical illusion," to set aside any miracle! Why, the man has only to say, "Nature," "law," "development," "endless series," "spontaneous generation" — such high-sounding words, liberally and discreetly used, are enough to explain away from such a man the finest natural proof of God that could possibly amaze a philosopher. Who needs be told it — while it is notorious that men can doubt anything they choose, and on occasion can persuade themselves that fields and floods, hills and heavens, are nothing but fictions of the brain! Yes, there must always be more or less haze between the sun and a marsh. If the marsh is sulphurous and hot, the haze will be a cloud through which shall not struggle the faintest outline of an orb, or even of a halo.

But this is, possibly and not improbably, but one fold of the veil which sin hangs between us and God. Were God from a certain point to increase the general evidence of Himself, He would be sure to increase the responsibility of every human being. But He would not be sure to increase the aggregate faith and virtue of the world. Indeed, would He not be sure to do the contrary? In some cases a change for the better would be produced: as certainly in many other instances the change would be for the worse. The increased light would be resisted — sometimes by unbelief and sometimes by believing unrighteousness — and consequently more guilt incurred and more damage sustained than under the old state of things. That there would be many on whom this sad effect would display itself, no one familiar with life will doubt: that the number of such persons would not be so great as to outweigh with their disasters all the good done, who has a right to affirm? Just bethink yourselves what multitudes daily manage to resist any amount of evidence when their inclinations, prejudices, habits, are against it! Just bethink yourselves what multitudes fail to walk by the evidence they accept; though accepted as proving at least the possible truth of so heavily sanctioned a system as the Christian religion! Beyond a certain point in the accumulation of evidence for a God, the damaged multitude would surely become a damaged majority. For, after a certain stage of revelation has been reached, it is no longer want of evidence that prevents faith, or want of motive that prevents virtue; it is disinclination and frowardness of heart. Increasing the amount of proof does not touch the seat of the trouble. Double the proof, treble it — it makes no difference; the men are unbelievers still. With the mind made up, it is just as easy to shut the eyes on a mountain as on a mote; just as easy to turn back on the north when it is flaming with ruined rainbows as when bare of a single ray. So there must be a point in the accumulation of evidence for a God, when, while increasing as fast as ever the responsibility of a wicked world, it ceases to increase its faith and virtue. Here a perfect God must break off His proofs. Sin has added another fold to its veil.

But is there not yet another fold from this source? The necessity of not doing more harm than good would limit the revelation to sinners: would not the necessity of doing them as much good as possible limit it still further? It was once said of a certain text-book in science that it was too simple and explanatory for the use of students in a college; it did not tax and discipline their minds sufficiently. No one would have denied that the book was a good one; that a very valuable culture was derived from the study of it; that its advantages greatly outweighed its disadvantages. But these facts did not establish the propriety of holding the algebra to its place in the college. The question to be answered was, not whether the treatise was on the whole useful, but whether it was as useful as some other would be that should throw the students more on their own resources. And this question was answered in the negative. It was decided that a less explanatory book would serve the purpose of mental education better. So Stanley was substituted for Day, as perhaps Day had been substituted for Euler. Now, is any one competent to say that God has not seen reason to make a similar decision in regard to what is the best mode of teaching men theology, the science of God? As a good Being, His object in revealing Himself to the world would not be merely a good moral discipline and culture, but the best possible. And it is conceivable that even as the obscurer manifestation of algebra may do the most for the crude and wayward intellect, so the obscurer manifestation of God may do the most for the crude and wayward heart. May it not be a noble discipline of character to inquire after God humbly, patiently, fervently, in the face of some difficulties? May it not be a noble discipline to practice carefulness, love of truth, fairness of mind, prayer for Divine guidance, as a requisite to success? And when success comes in this travailing way, would not it and its results be all the more highly prized for the pains taken? These are suggestive queries. And they make it unsafe for any one to declare it improbable that the veil which sin hangs between us and God is not thick with the necessity of securing the best as well as a good moral training for a depraved world.

So add a threefold veil of sin to those four veils which, from the nature of the case, must interpose obscurity between us and God. He neither wants to turn the world into an Olympus, nor into a Necropolis, nor into a Bedlam, nor into a Bœotia, nor into a Stonehenge, nor into a Pandemonium. An unstinted revelation of God — the man who asks it knows not what he asks! Does he want the creature to take on the full stature of the Creator? Does he want such an exhibition as, with its terrible sheen and trumpet, shall blast his life away?

Does he want Reason to start headlong from her throne, and maunder out of the dust? Does he want the faculty of astonishment calloused by daily violence into an Ironsides which no marvel, though catapult-hurled, can impress? Does he want men made with stocks and stones for souls, instead of free moral natures; or that God should content Himself with something less than the good, or even the best, in His method of dealing with sinners? If not, he must be content to have the world look toward God with at least five veils dimming His majesty and existence.

At least this number of veils may or must depend between a God and this sinful world. And now the question is, whether they will account for as much obscurity on the Divine existence and majesty as an objector may properly assume to exist. How much obscurity is that? How much will intelligent theists admit? They will admit that some persons have no faith in God; that multitudes have far less faith than would be desirable; that still greater multitudes have an idea of God that is troublesomely weak, unimpressive, and uninfluential. This is the obscurity we confess to exist within men. As to the obscurity without them, we confess something and claim more. We confess, of course, that the revelation of God is not overwhelming in the amount and quality of its evidence, so as to make unbelief and negligence impossible to such beings as men; nor do we know of any evidence in earth or heaven, within the whole realm of even ocular and mathematical demonstra tion, that does possess such a character. We confess that the evidence is not such as to totally exempt men from care and pains in order to receive and retain its full force; and we have yet to learn that it would be desirable to have it so. Thus far we confess. On the other hand we claim, and hold ourselves ready to prove, that the existence and majesty of God are supported by evidence that is decisive; evidence that is sufficient; evidence that is very great; evidence that is greater by far than upholds any other moral thing; evidence fully as great as men seem disposed to improve; evidence great enough to secure, in every age, almost universal faith in at least one Worshipful Intelligence indefinitely superior to man in wisdom and power, and quite universal conviction of the possibility of such a Being (which, so far as the practical guidance of life is concerned, is almost as exacting a principle as faith itself); indeed, evidence great enough to give moral certainty and a renovated character to every person from sunrising to sunrising who will use it faithfully. We claim, and hold ourselves ready to prove, that Nature and the Supernatural have not spared themselves; that they are generous witnesses for God; that they vie with each other in the richness of their testimony; that God has a shrine in every history, a temple in every science, a stoled priest with his Novum Organon in every bosom; that the wide campus of matter swept by microscope and telescope as far as yonder picket nebula is everywhere covered with hieroglyphics of Him which no Champollion is needed to decipher, everywhere hung with His cartouches and coats of arms which no college of heralds is needed to explain, everywhere tracked with His giant foot-prints vastly more scientifically intelligible than any of these fossil scriptures of the Connecticut which so nobly enrich your museum — further and chiefly, that the supernatural evidence carries itself still more regally in a God who has often spoken audibly with men; has often stood among them in visible personal forms; has dwelt for thirty-three years on the amazed and panting planet in a human body; has maintained for ages an oracle whose Delphos and Dodona shone with miraculous Shekinahs and infallible Urim and Thummim; has made the future visible, the dead to live, the earth to tremble, the heavens to blaze, the angels to fly singly or in armies along the sky in attestation of Himself; indeed, has even personally come down in presence of forewarned and expecting hosts, embosomed in a storm of miracles and with ineffable pomp, as if to silence, once for all, the clamors of such men as say, "Nay, father Abraham, but if one went to them from the dead, they would repent;" and, finally, has scattered these direct manifestations and these attesting miracles, with their diaconate of special providences, through the ages as liberally as can be shown consistent with their best effect, bridging the intervals between them with well-accredited and well-believed traditions from amid whose mighty arches and colonnades and picturing perspectives they can, not improbably, be seen to better advantage by the majority of mankind than from the portico itself.

Lo a Man of whom I have great things to say! He had profound faith in God. He not only believed in the Divine existence and perfection and government; but that glorious idea seemed to him very much as did the ground which sustained him, the air which he breathed, and the heaven which rained on him from its azure cope the glory of sun and stars. — Is God all about me? Does He look and work on my right hand and on my left, upon me and within me? Do I never go abroad but that His providence paces along; never rest at home but that His sleepless sovereignty watches at my bedside? While I am thinking, is He busy among the thickly coming fancies and arguments; prompting, repressing, proportioning with a tireless hand? While I am speaking, is there no slightest tone that does not reach His quick ear; and no hearer's heart into which He is not looking, and where He is not working in behalf of the truth? The wind that sighs under these eaves — is it true that the pulse of almighty power is in it? The cloud that sails yonder — is it so that an omniscient thought is riding on it, hither and thither, for its secret mission? Each ray of light that makes its way through these windows — is it feathered with a Divine purpose, and is every minute reflection from wall and seat and dust-particle presided over and governed by a single personal agency as real as that which turns over the leaves of this manuscript? — So thought that man of faith. So was he convinced. So, indeed, he almost seemed to see. Other eyes than those of his body seemed to dwell behind and look through those grosser orbs, and to see things too subtle and essential for them. The common world lay insphered in a supernatural. All things were "living and moving and having their being in God" — as says Euripides, Ω γῆs οχημα, κἀπὶ γῆs ἔχων ἕδραν — Chariot of the world and having His throne above it. In the greatest national affairs, and in the obscurest domestic history as well, played a Divine hand. Whatever befell, whether sad or glad, was the providence of God. The fields of earth and air and sea, instead of appearing as so many platforms on which the machinery of natural causes was milling out science and inevitable results, seemed so many Fields of the Cloth of Gold on which the Great Supernatural was accomplishing in his own proper person the grand part of King of Nature. As to him of Patmos, a shining Infinite Presence seemed moving about among all those ages and empires and ecclesiasticisms and homes and individual joys, sorrows, virtues, sins — above all, mingling freely with all his own affairs, and pacing with unwearied step throughout body and soul and all personal mysteries. Not more real was yon mountain that half shuts out the day. Not more real was yon ocean that awes so many lands with its lordly voice. So temptations were cobwebs to him. So dangers were no dangers. So trials were a heavenly discipline. So life and death were only different ways of spelling the same word. Reproaches, ill-fame, martyrdoms — what recked he even for these? He endured as seeing Him who is invisible. And all skepticisms, and philosophies falsely so called, though pretentious and crested as ever were ocean waves, broke harmlessly upon him. If at some point that rooted continent occasionally seemed to lose a few grains of its rocky substance, it was only to more than add them to some other part of its great coast. So all the while the Terra Firma rose and grew and peopled itself — as America is now doing amid the buffets of two oceans. It was a grand sight to see — that great, immovable, progressive faith; and the profuse billows of French unbelief breaking into smoke or ever they touched its mighty sides!

How came this? Was the man intelligent? He was a philosopher by nature and training, and such an effulgent thinker had not been seen for many a day. Was he practical? No man managed his common affairs with more sobriety and discretion. Was he a victim of the unaccountable phantasms of youth or of age? He was at the noon of his great and well-balanced faculties. And yet what a triumphant believer! How shall we account for him? I will tell you. Years ago that man was wholly without faith. Somehow his skeptical reading and his puzzling speculation had by degrees taken away the God of his fathers. Look where he might, there were no satisfactory signs of the supernatural. He could see nothing but great Nature. And from out his logical fogs he looked with mingled pity and contempt on men simple enough to believe. But, as said Plato, without a God no man is at rest. So one day he caught the light from a new angle. He was startled. An immense Perhaps stared him in the face. He became at once an earnest inquirer. He set himself to examine the fundamental religious question with as much pains as reasonable men use in their very important secular concerns. He thought that much was plainly reasonable. He also thought it reasonable to invoke the God who might be, and who, if real, could easily and abundantly help. So he called freely on the possible Divinity — as a man lost in the forest will sometimes lift up his voice in loud calls for helpers whom he can neither see nor hear, but who, for all that, may be near at hand. — Another thing he did meanwhile. He freely consulted that Book which above all others credibly purported to be a Divine message. He found its aroma very peculiar. It was not that of materialism. He found its ways not like those of man. Nothing like them in science or history or business. Every verse had its face turned God-ward. Every chapter stood gazing upward, like the Christian apostles on Olivet. Every section cried God — "Him first, Him last, Him middle, Him without end." The whole credible message gravitated, pointed, and prayed toward this great Center. Gradually the habit of the Book became the habit of the man. Gradually, as he read, he felt himself lifted into purer airs and clearer prospects. Gradually, as he read, he felt a new sight quickening and reaching forth from the depths of the old — especially as he went on to test the message, according to its own invitation, by a personal experiment on its adaptations to human wants and the faithfulness of its promises. — And so he began to live in such a way as made the thought of a righteous God full pleasant. Then his faith ripened fast. The vintage grew heavy and purple every day beneath beams so bright and genial. And at last the man came to have, if not a Divinity, at least a Divine work going forward bravely within him. Great reforms took place. Great reconstructions of character occurred. A structure far nobler than the proud baronial halls in which he dwelt, arose within him. And, withal, his soul with its real though spiritual ear caught the sound of great spiritual processes of rectification and repair and cleansing and enlargement going on within its chambers, and became conscious of being touched and lifted and wrought by a wondrous agency not of this world. — When the builder is doing little or no work in my house, I may sit quietly, and quietly look out of the window, and scarcely ever think of there being such a person, though he is busy through the village from morning to night. But let him come into my own dwelling, and begin extensive repairs — let him drive the axe and the plane and the hammer with the energy of a strong man in the very room I occupy — let him lever and screw up the whole building to rest on a new and higher foundation — let him take out old and decayed timbers and replace them by new ones, dig down my walls of plaster and panel others for me in immortal oak, put up addition after addition, and so go on with infinite hewing and carving and beating to transform my hut into a palace — can I help realizing that builder then! He will scarcely ever be out of my thoughts. When my eye does not actually rest upon him, there will still be in my mind a vivid picture of his doings, as matters very real and very near — especially if they are never carried on independently of me, but with hourly reference to my judgment and assent. — So it was with that great believer. As soon as a thorough moral transformation came to be fully in process within him, he became aware of a Divine Builder as never before. Spiritual senses asserted themselves. There were spiritual sounds and motions and touches and thrills which could not be mistaken. The chambers were perfumed with a heavenly breath. The corridors echoed to a heavenly step. A heavenly voice sweetly rang through the vaulted halls. He felt that he was experiencing God. *Communion* with God was established — lo, it had been said, "We will come unto him and make our abode with him." So while God was building up the character, He was just as fast building up the faith, and fulfilling the promise that whoever does His will shall know of the doctrine. Thus it was the man became continental in faith. Thus did he reach moral certainty. Thus did he see his cornucopia filled to overflowing, and passed almost beyond the power of understanding how men could think of complaining of the obscurity of God. And thus it came to pass that when he died, it was with a far more confident expectation of waking on God than he ever had, when falling asleep, of opening his eyes the next morning on his own ducal domains of Broglie.

Such was the method by which a seeing faith came to him. And it has come in the same way to many another. I take it on me to affirm that it might come in the same way to all — to all these men who are complaining of the scant light. Let them try this specific. Let them try it, if they would have moral certainty on the most important of all questions. They cannot claim that the method is not reasonable, plausible, and essentially philosophic — if there is a God. It is such as a God would not be unlikely to put men upon. And it is fortified by an experience that deserves to be called scientific. It is a principle of science, carrying with it the universal suffrage of modern scientific practice, that any objection to an hypothesis is sufficiently met when a simple, natural, and perfectly credible way of solving it, in consistency with that hypothesis, can be stated. But in this case we have more than such a natural statement. We have that supported by a wide experience and induction. Let every believer in experimental and inductive science take notice — until he also is able to join that unsandaled and elect company which in every age has not failed to look upward and around with awe-stricken faces, and to softly say with the supreme confidence of sight these purifying words, "Thou compassest my path, Thou besettest me behind and before. Whither shall I go from Thy Spirit!"

Does any one imagine that such obscurity on the being and majesty of God as may belong to such a system of revelation as this, is not sufficiently accounted for by our five veils — one of them, at least, of indefinite thickness? I hold that the situation of mankind in respect to Theistic light is very like our situation in respect to sunlight on some fair day of summer. There are floods of rays abroad. High hill-top and profound valley shine. But some men are in less sunny places than others, some look forth from northern aspects, some live in smoky Birminghams, some are inclosed in curtained and shuttered houses, some are busy in sewers and pits and coal-caves, and some are blind. "Why is it so dark?" say some of these men. "What darkness do you mean?" I ask. "Do you mean the darkness in your cloistered houses, your fuming furnaces, your subterranean dens? Why, your shutters and smoke and ceiling of opaque earth-strata sufficiently account for that. Ascend from your pit; go forth a few miles from your smoky Birmingham; make your shutters and curtains describe a full semicircle in favor of the day, and you will find a wonderful improvement in the brightness and salubrity of your surroundings. No Leslian nor Wollaston gauge will be needed to ascertain it. But perhaps you mean a darkness on the general face of Nature? If so, then I have to say that I have not discovered any particular scarcity of light there — there seems to me enough for all practical purposes; indeed, it seems to me that the amount of light is exceedingly great. Still I am willing to admit that it is not as great as might be. Were the earth to cease reeking its vapors; were the cloddy fields to change into pure and burnished gold; were each slant ray to become a perpetual perpendicular; especially were the sun itself to draw nigh till all heliometers are abashed and the whole sky is filled with its flaming disk, we should have still more light: the want of these things is so many veils before the majesty of the sun, and all the dimness — if you please to call it such — which you perceive they will sufficiently account for. Would you have these veils removed? Then prepare to perish in the blaze; or to part with senses and reason in the appalling effulgence; or to have all objects reduced to the same dead level of commonness by the undiscriminating and perpetual dazzle; or, finally, to become less vigorous, manly, virtuous persons than you now are — perhaps like yonder enervate and voluptuous tropical Asiatic, sweltering alike in his sun and his sin — do this, or totally change your natures. With such natures and tendencies as you have, I think that this bright temperate zone and this golden sun of a half-degree diameter is the very best thing for you — wonderfully better than a sun whose fiery shield fills the whole astonished one hundred and eighty degrees. Who has a right to deny it? And pray, who has a right to deny that this bright though tempered revelation of God which theists claim, with men only held responsible for such measure of light as they have, is the very best revelation which could possibly be furnished to fallen beings!

VI.

HARMONIES WITH NATURE.

Αρατε εἰς τὸν οὐρανὸν τους ὀφθαλμοὺς, καὶ ἐμβλέψατε εἰς τὴν γῆν κάτω.

Ζεὺς, Πυθμὴν γαίης τε καὶ οὐρανοῦ ἀστερόεντος.

Orpheus.

VI. Harmonies with Nature.

SIXTH LECTURE.

HARMONIES.

IN the last lecture I mentioned some sources of obscurity which must embarrass a manifestation of God; namely, the narrowness of our intelligence, the frailty of our bodies, the frailty of our reasons, the frailty of our sensibility, and the sinfulness of our hearts; and then endeavored to show that these will easily account for all the dimness of Divine manifestation that intelligent theists will admit to exist. Their admissions will be scanty. It seems to them that the light is that of broad summer day — admirably broad and brilliant, though capable of being suppressed to any extent by the personal habits of mankind.

This concludes what I have to say in answer to objections to the doctrine of a God. These objections are few. Atheists are more accustomed to assail the proof of the doctrine than the doctrine itself.

In coming to the positive evidences for the Divine existence, we are met by the fact that theists and atheists agree as to the advantage of approaching the question of a Divine Being with a mind freshly steeped in the leading facts and courses of nature. The atheist claims that nature makes on minds thoroughly imbued with her spirit an impression adverse to faith; and points in evidence to some eminent cultivators of the physical sciences who have been as skeptical as they have been scientific. So he is in favor of the study of nature. On the other hand, the theist is in favor of it for the very opposite reason. He denies the atheism of science. He refuses to infer it from the unbelief of some French and German philosophers — with here and there a second-rate English disciple — whose minds from childhood have been poisoned with the writings of Voltaire and his school, who have seen around them only a grotesquely corrupted form of religion, and whose private lives for the most part were such as to make it greatly for their interest to have no God. To him the case of such exceptional men only shows the exceeding force of native depravity, evil training, evil surroundings, and evil habits, at withstanding the natural tendency of their pursuits. This tendency he regards as strongly theistic. He thinks he sees premonitions, prophecies, presumptions, and even proofs of Divinity in the great universe that expands around him; and believes that, other things being equal, the more fully one comes under the influence of the astronomy, the geology, and the other branches of natural science whose findings have amazed mankind, the more easily he will admit and the more strongly he will hold, the doctrine of a Divine Being.

What all classes think it well to do, let us attempt. We will attempt to place our hearts still more fully *en rapport* with nature. We will, if possible, get them into yet closer communication and sympathy with its great leading facts and courses. These are chiefly astronomical. Yet I shall not restrict myself to astronomical facts, technically so called, but shall allow myself to gather from the whole of that broad field of science of which astronomy is the undisputed and all-comprehending Chief. And I can not but think that the effect will be to preclude objections, to furnish presumptions, and generally to dispose the mind to a mighty faith in God. I am persuaded that any man who can be fairly set down in the midst of nature, and thrown honestly open to all its subtle inductions, magnetisms, inspirations, will silently drink in theism, as a fleece spread out under the stars drinks in the dew.

Suppose it claimed that a certain veiled painting is the work of Titian. If, on gradually lifting the veil, we find exclusively trait after trait such as might have been expected in a work by that great master, our disposition to think favorably of the claim increases with every step: and if, when the canvas is entirely exposed, every leading feature seems Titianic and the whole worthy of such an author, our minds are far advanced toward faith — they are in a state of high preparation for any ulterior evidence, and only comparatively little of it will be required to secure full conviction. And this is reasonable. Previous to examination, how could we be sure that there were not lurking under that veil incompatibilities, or at least disagreements? Now our uncertainty is removed. We have found positive harmonies. The facts match the claim. The picture is such as might have been expected from Titian — such indeed as he would surely have painted. His great characteristics are strikingly here. And these are so many verisimilitudes, so many presumptions in favor of the claim: and, in the absence of all evidence to the contrary, at least authorize the critic to stand at the very verge of assent, facing it kindly and with foot uplifted, ready to cross the border at the first competent invitation. Let such an invitation come in the shape of an assurance that the painting is almost universally accepted as the work of Titian, especially among the most intelligent and fair-minded judges; further, that the hypothesis which ascribes the work to him is, as compared with other hypotheses, altogether the simplest, the least embarrassed, the most useful, as well as the most historical — this would and should plant his feet in the very center of faith.

Now it is claimed that nature is the work of God. Let us, step by step, unveil its leading features and see if they do not strikingly harmonize with the claim: and, as they may be found to do so, let unbelief approach its frontier; and, when at last the general scheme of nature appears characteristic and worthy of God, let the traveler at least stand on the last boundary of his chill and somber territory, all ready to cross with decisive and ringing step into a brighter land at the first summons of the positive evidence.

What I propose, then, in the present lecture, is to illustrate the general harmony between nature and the doctrine of a God. Of course, a few specimen illustrations are all that can be offered. One will do well to feel the pulse of nature still more fully in the works of Ray and Good and Paley, in the Bridgewater Treatises, and in later works of the same character.

One of the most striking features of what we call Nature is its vastness.

I do not forget that I am speaking to those who have become familiar with the wonders of physical science. But neither do I forget that even the scholar must refresh his impressions of things in very much the same way with other men. So I ask you to think of plains stretching to the horizon; of mountains piercing the clouds; of roomy continents anchored in roomier oceans; of this whole earth-sphere, with its huge baldric of twenty-five thousand miles, covered with innumerable vegetable products, peopled with men to the potential figure of a thousand millions, swarming still more potentially with the lower animals, and so flooded with microscopic life that almost every cubic inch of air and water and soil is panting with an incalculable population, — some of whose smaller individuals multiply themselves into one hundred and seventy billions in four days; gather their five hundred millions in a single drop of water; and yet make up, with the stony cerements of the merest fraction of their fossil ancestry, whole mountains and geologic beds. Such is our world. Out in yonder vault, find that millionfold world which we call the sun, with its invisible retinue of a hundred earths; out in yonder vault, when night falls, find a thousand suns similarly attended; with tube Galilean, thousands more; with tube Herschelian, millions more; with tube Rossian, billions more. Is this the end? What astronomer for one moment imagines that another enlargement of the great speculum at Parsonstown would show our vision to be already hard up against the frontiers of nature? Not even Darwin doubts that successive improvements in the space-penetrating power of our instruments would go on indefinitely opening up firmaments at every step. Where is the verge of the universe? Who would undertake the roll-call of its orbs? Who dares to say that he could count through the grand total of its firmaments, even though he should count a thousand years? Figures go but a small way toward expressing the dimensions of such a universe — whether one considers the number of its worlds, or the expanse of space through which they are distributed. Our world spins round its ellipse, of well-nigh two hundred million axis, without ever having a neighbor nearer than thirty millions of miles, save its own moon. The interval between our sun and the nearest star of the same galactic nebula is twelve hundred thousand times this distance. And then the distance from nebula to nebula — it is absolutely awful. Our telescopes sweep a sphere of stars whose diameter is seven millions of years, as light travels. Calculation covers its abashed face with its great wings in the presence of these overwhelming amplitudes. And such is nature!

Certainly such a universe as this does not cry out *against* the existence of a God whose essential attribute is immensity. On the contrary, it is just such a universe as one would have *expected* to come from such a being. Nay, given a Deity who is practically at home in every point of space, whose attributes are laid out on a scale of unbounded vastness, to whom it is just as easy to make and govern a trillion of worlds as it is a grain of sand, and the imperial fitness of things would *demand* that he people vacancy with very much that profusion and breadth of being that we actually see. The work ought to express and honor the workman. And when I am *told* of an author of nature who is immense with a threefold boundlessness of intelligence, might, and years; so that to him our great and small, our far and near, our center and circumference — though that circumference sweep around all the expanses of modern astronomy — are practically the same; so that he can properly challenge, "Do not I fill heaven and earth?" — when I am told of this, and I then place myself out under the open dome of nature, amid its exuberant objects and marvelous stretches, I feel myself silently drinking in predispositions to faith as the fleece spread out under the open heaven drinks in the dew. I feel that the doctrine matches facts; that the theory has in its favor a comprehensive verisimilitude and presumption; that Nature, instead of saying, "There is no immense God," significantly asks, in a tone of encouragement and with a look of incipient expectation, "Is there *not* such a Being?" In fine, I feel that our slight lifting of the veil from the painting has disclosed a feature strikingly characteristic of the great master to whom the work is attributed — a feature which, in the absence of all counter-evidence, naturally sets our faces faithward — one, of several harmonies which, as successively presented, will warrant us in looking faithward with evergrowing kindliness of aspect.

Notice with me the *variety in unity* that characterizes Nature.

Some hundreds of millions of creatures on our earth are so much alike that we put them into a class by themselves and call them men. They are all alike in certain fundamental features; and yet each man differs materially, both in body and soul, from every other man. So of every other class of things — animal, vegetable, inorganic; while there is a sub-stratum of unity among the members of each, on account of which they are classed together, there is not one which is not very unlike, in many respects, all its fellows. All animals have great points in common: but how many, many sorts of animals; and how great the difference between the eagle and the microscopic mote, between the cetus and the polyp, between the most perfect man (body and soul) and the rudest of the polypi! All vegetables are similarly constituted: but whose memory can master all the distinct kinds of vegetables in the wide interval between the spire of grass and the huge tree that wrestles victoriously with stormy centuries; and reckon up the great differences that exist, as to shape and size and color and flavor and odor, among fruits and flowers and leaves and grasses and shrubs and trees. Great threads of unity obviously connect all the forms of terrestrial being, organic and inorganic; but this we know, that, if only single specimens of all the plainly separated species were attempted to be brought together into one Crystal Palace of a museum, we should have to roof in empires, instead of acres, in order to accommodate their mighty array: and as our eye would run over the whole superb collection, and at last bring together the two termini — viz., the material man and the material stone just crumbling into dust — our sense would be that of a miraculous diversity efflorescing out of the unity of our world. So with those other worlds that shine or hide in the vault above. They are all spheres, all have orbitual and probably axial motions, all are governed by the same principle and law of gravitation, all are lighted and colored and warmed by the same mysterious element or impulse; but on such basal unity is superimposed an almost infinite variety. Observe our solar system. One member of it is self-luminous, and, relatively to the other members, a nearly stationary body; the others are dark, and far-wandering planets. One is one hundred miles in diameter, another nearly one hundred thousand, while still another contains more than eight hundred times as much matter as all the remainder of the system can boast. Some have atmospheres and seas, others have neither. Some have moons, others have none. Saturn rides forth in the pomp of three great equatorial rings, as well as of eight moons; no other planet is similarly furnished. These orbs of our system differ greatly in density — one is as lead, another as cork, another still is mere vapor. One receives seven times as much light from the sun as we, another only a three hundred and sixtieth part of as much. Neptune's year is equal to one hundred and sixty-five of our years. Saturn's day is only one-half of our day. Of course the products and scenery of these worlds, as well as the constitution of their inhabitants, must differ exceedingly. But pass we on to the region of the fixed stars. Have we escaped into immeasurable uniformity out of immeasurable variety? Lo! we skirt systems, clusters, firmaments, and never two alike, while some stand apart by whole universes of difference! Lo, systems with several suns each, from one to a hundred! Lo, systems lighted, some with white suns, some with ruby, some with emerald, and some with suns of many different colors! Lo, suns differing exceedingly in size and amount of light they shed: for the great Sirius that flashes first magnitudes on all our charts as well as on the dazzled retina of the savage, is not as near to us as the little 61 Cygni, and its light must be equal to that of two hundred and fifty suns like our own! Alcyone shines with a force of twelve thousand suns. And then we have suns themselves combined into systems of all sizes and shapes — systems of two, of three, of many, of millions, — firmaments which, under the name of nebulæ, are the last generalization and most stupendous variety of modern discovery: sometimes rolled up into spheres; sometimes gathered into circular or elliptic rings; now fan-shaped; now like an hour-glass; now broad wheels of compacted suns, large, glittering, and sublime enough to under-roll the chariot of immeasurable God. There are not two leaves or grass-blades perfectly alike in all this verdant world; not two worlds, nor systems of worlds, accurately alike in all the prodigious realms of astronomy.

Now no one, to say the least, can claim that this vast variety imbosomed in unity makes positively *against* the idea of one Creator of boundless invention and executive faculty. On the contrary, it is just what we should have *expected* from such a being. Given just such a many-sided, versatile, complete Deity as is affirmed — we should say that, in case he should set himself to produce a vast universe, he would be *likely* to produce one in which great outlines of unity would be steeped in immeasurable variation; one in which resemblance and diversity, both robed and featured like goddesses, would hold each other by the hand and go treading with wedded and festival step up and down the whole quickened area. Nay, this sort of universe one would make *sure* of finding; would be greatly disappointed if he should not find. The eternal laws of his own nature would demand it of the Great Builder. The invincible beauty and fitness of things would demand it. Perfect uniformity, however piled up in magnificent magnitudes — even a uniformity only varied after so cramped and frugal a fashion as would be perpetually suggesting poverty of resources — would belie the inexhaustible Divinity. If he build at all, he must not misrepresent and disparage himself in his work; his fruitful nature, teeming with all imaginable fertilities and seeds, must surely blossom into very much that marvelous fruitfulness of product and pattern which we observe. And when I am *told* of an author of nature whose being swarms in resistless force toward every point of the compass, nay of the sphere; who is both a unit and a polygon, facing every desideratum and possibility with a flashing side, both of thought and action, that out-dazzles the sun — when I am told that such a being is the author of nature, and I then put myself forth under the open dome amid the glorious diversities that root themselves in the glorious unity of nature, and open myself freely to all their subtle suggestions and magnetisms, I feel myself drinking in predispositions to faith, as the exposed fleece drinks in the dew. I feel that again the doctrine matches facts, that again the theory has a comprehensive verisimilitude and presumption, that Nature instead of saying that there is no God whose unity is arborescent with endless varieties of beauty and power, significantly asks, "Is there *not* such a Being? In fine, I feel that our continued lifting of the veil from the painting has disclosed a second trait strikingly characteristic of the Great Master to whom the work is attributed; a trait which, added to the first, warrants our faithward look in taking on new kindliness of aspect.

Another characteristic of nature deserving of notice is the *perfection of its details.*

The exquisite finish of nature in its minutest parts is about as wonderful as its vastness and variety. Scan that leaf. Examine the wing of that butterfly. Let the tinted and polished antennæ of that moth glitter in the focus of your instrument. Subject to the skilfullest notice of science and art the smallest veins of any animal or vegetable. Push the analysis just as far as possible, and submit that last visible minimum of organization in the crystalline lens of the cod, with its five millions of muscles and sixty thousand millions of teeth, to the most searching criticism of the superbest microscope. What exquisite details! What elaborate refinement of workmanship! It is not as with some master-piece of human painting — the main points only cared for, while all the subordinate are too rude to bear close inspection. Titian painted this landscape. Well, it is worthy of him — the general effect is beautiful. Yet, if you approach, and closely examine the foliage of the trees, the grass with which the canvas is green, or even the limbs and features of the animals, they will be found very coarsely and incorrectly executed. The microscope turns the most finished work of man into coarseness and clumsiness — indeed, almost immediately carries the sight where traces of skill have totally disappeared. Not so with the works of nature. A real landscape you may analyze to your heart's content, and inspect its details as critically as eye armored with lens can do, without finding the workmanship growing less exquisite the further you push inquiry. A real man — you may descend to the minutest particulars of his organization, and get as near its primary elements as an Ehrenberg with his superb instruments and practiced vision can carry you, without finding the least falling off from that delicacy of execution which appears on the larger masses and outlines of the body. So everywhere among natural objects — the great and the small, the outlines and the minute filling-up, as far as utmost optical resources can carry our observation, are wrought with apparently the same overflowing outlay of attention and skill. It is not so in a few instances merely, nor in a thousand — it is so universally.

That there are any so preposterous as to think that this feature of nature makes positively *against* the idea of a sparrow-watching, hair-numbering, and thought-weighing God is, of course, not to be imagined. Of course, it is a feature that fully *harmonizes* with such an idea. A nature finished exquisitely down to the most infinitesimal of its details is just what one would have *predicted* from a God of this description. Announced the fact that He was about to create, and expectation would have stood on tiptoe to look for just such a nature as we see. A God for whose vision nothing is too small, who necessarily gives as complete attention to the affairs of an atom as to those of an empire, who can concentrate his almightiness with as much freedom and accuracy on a mathematical point as on a world, who is embarrassed no more by unlimited multiplicity than by unlimited minuteness of details, who can with equal ease paint a landscape on the point of a needle — say, if you please, forty thousand of such landscapes at once, with all their innumerable and minima particulars, back of the reticulated eyes of a single butterfly — can with equal ease do this, and roll a solar system on its triumphant path about the Pleiades ; do I not know that a being with such a striking attribute as this would surely give it expression in his works ? Do I not know that he who is equally at home in maxima and minima, and to whom beauties and glories in the world of infinitesimals would be just as apparent and practicable as they are in the world of infinites, would lay himself out on the one very much as on the other — would effulge himself into the microcosmos very much as into the cosmos? When, then, I am *told* that such a being is the author of nature, and I proceed to place myself out under the open dome amid the exquisite elaborations that swarm on every hand down through the veriest miracles of littleness and detail, and to uncover myself candidly to all their subtle whisperings and magnetisms, I feel myself softly drinking in predispositions to faith, as the exposed fleece drinks in the dew, I so feel the force of a doctrine matching facts, and buttressing itself again and again with comprehensive verisimilitudes and presumptions, that to me nature becomes articulate, and, instead of swearing with uplifted hand that there is *no* wondrous God, significantly points upward, and, with bated breath and expectant look, asks, "Is there *not* such a Being?" — in fine, I feel that our continued lifting of the veil from the painting has disclosed another characteristic of the Great Master to whom the work is attributed, the third of those several harmonies which, as successively presented, warrant us in looking faithward with ever-growing kindliness of aspect.

Another feature of Nature is what I shall call its *wisdom*.

The world is full of what, if accepted as the work of an intelligent being, would be called contrivances — adaptations of means to ends — often of the most complex and elaborate description.

For example, the birds — how admirably adapted to flying; in shape, feathers, bones, wings! The fishes — how adapted to swimming and life in the water; witness their shape, their smooth and unctuous scales, their pairs of fins, their tails and gills! The land-animals — how adapted to walking and running and feeding on the earth's surface; to eat the grass or catch their special prey! The trees — how adapted to stand firmly; by their roots, their perpendicularity, their balanced branches, their moderate flexibility — how adapted for shade, for abating the violence of winds, for fuel! Or, if you will consider particular organs of the organic tribes, look at the bark of trees as related to their nourishment, at the web-foot in its double relation to land and water, at the teeth and other preparers of food for the stomach, at the stomach as a preparer of food for the blood, at the lungs as purifiers of the blood, at the heart as the engine for forcing the blood to all parts of the system, at the hand as the general servant of the whole body; in short, at almost any organ of either animal or vegetable structures. The adaptations are wonderful. They are physical miracles — the means are shaped and applied to the ends so exactly, beautifully, triumphantly. For example, no work of human ingenuity that ever you saw is equal to that natural marvel, the human eye — an organ having reference to an element quite external to itself, whose chief source is millions of leagues distant; and also to millions of external objects which compose our scenery of earth and sky — an organ placed in the most elevated part of the body so as to command the most extensive prospect; placed in the front so as most readily to preside over the direction in which we habitually move; placed in a strong bony socket which defends it from the heavier external injuries; imbedded in a soft cushion, so that its delicate texture can not be hurt by the bony walls around it, as it rests on them, and turns swiftly hither and thither at the bidding of the will; furnished with lids, like curtains, to close over it in sleep, to wipe it, to cut off the outer rays of light that would confuse vision, to protect it by their involuntary and instantaneous shutting against the lighter kind of injuries; furnished with an apparatus of muscles by which it can be rapidly turned at choice in any direction, so as to vary the field of vision as the needs of life may suggest; furnished with a self-acting system of appliances by which the ball is kept lubricated for easy movement; furnished with a conduit to carry off the superfluous moisture; furnished with just that shape, out of ten thousand possible shapes, which mathematicians have demonstrated to be the only one which can refract all the rays of light to a single surface, and thus afford distinct vision, viz., that of an ellipsoid of revolution; furnished with a retina or natural canvas on which its pictures of external objects can be formed, of just the right size, and at just the right distance behind the lenses of the eye; furnished with lenses of different substances having different refractive powers, thereby preventing the light from being resolved into the prismatic colors, and thus misrepresenting and uniforming objects; furnished in front with a perforated membrane that by self-adjustment adapts it to different degrees of light, also with a system of pulleys and ligaments that at a moment's warning alter its convexity and the relative position of parts so as to adapt it to objects at different distances and, what is more wonderful than all, provided in some inscrutable manner with the means of expressing the mind itself, so that one may look into its crystal depths and see intellectuality and scorn and wrath and love, and almost every spiritual state and action. Now, if this is not an amazing congeries of adaptations, there is and can be nothing amazing. If found to be the work of a human artist, it would be called a perfect marvel of ingenuity and wisdom. And yet some insects have twenty thousand such eyes combined into one. But the eye is only one among an infinity of natural contrivances. Animate and inanimate nature is mountainous and glittering with them. Down into the regions of the infinitely small, whither only the most searching microscopes carry the sight; up into the regions of the infinitely large and far, whither only mightiest telescopes lift our struggling knowledge; among the mech anisms of the atomic nations that people a single leaf, and among the mechanisms of those swarming celestial empires whose starry banners sweep our nightly skies — it is everywhere the same; exquisite adaptations crowding exquisite adaptations, profound contrivances (so inventors and mechanicians would be tempted to call them) heaped on profound contrivances, in such endless amounts and varieties of wise structure, as exhausts all human understanding and dwarfs into nothingness all the products of human ingenuity.

Does such a nature as this swear *against* a Divine Contriver. Does it protest against him, or testify against him, or breathe even a suspicion against him? Many absurd things are done in the world: but it will be hard to find the man who will care to deny the positive and emphatic *harmony* between the doctrine of an omniscient and omnipotent God and a universe crowded with such splendors of natural mechanics. A God of endless invention, and whose powerful and skilled hands can magnificently realize all that he has magnificently planned — we should *expect* that such a being, in case he should create a nature, would set it all ablaze with the monuments of his supreme intelligence and power — should be *disappointed* to find no such monuments, but, in their stead, mere stupidity or tameness of work. We should call the work unworthy of the workman. Nay, we should hasten to say to ourselves that we must have mistaken him — He could really be nothing more than such a petty divinity as the poor heathen have fabled to themselves. For we should be sure that one having unlimited command of ways and means, both as a knower and worker, would display it in his works. It being just as easy for him to have exquisite adaptations, and a gloriously endless variety of them, as to have no adaptations at all — it is plain what sort of nature he ought to make and would make. Now let me be *told* of a framer of nature in whom are hid all the treasures of wisdom and knowledge, whose light has in it no darkness at all, whose smallest deeds have from the hoary everlasting been pavilioned and charioted toward being amid the glories of Almighty Omniscience; and I then place myself out under the open dome mid the wilderness of wonderful constructions and chemistries, and candidly uncover myself to all their subtle sympathies and magnetisms — I feel myself, all silently, drinking in predispositions to faith, as the exposed fleece drinks in the dew. I feel that the God who is affirmed is just the God to match the nature which I see — here the ball and there the socket, here the foot Titanic and there its footprint, here the shapely hand and there its glove, here the sovereign sword and there the golden scabbard that just fits it — that these noble adaptations and mechanisms, spangling and blazoning all the fields of matter, are in rejoicing sympathy with the idea of a Creator who is wonderful in counsel and excellent in working; that the alabaster-box of precious wisdom that has been emptied, not only on the queenly head and shining tresses of Nature but on her very feet, scents bravely of One who is himself a "mountain of such spikenard;" that, in fact, the theory is again smiled upon by a comprehensive verisimilitude and presumption; that Nature, instead of swearing with uplifted hand that there is no All-wise Creator, with flushed cheek and upward-glancing eye of expectation, significantly asks, "Is there *not* such a Being?" In fine, I feel that our continued lifting of the veil from the painting has disclosed still another characteristic of the Great Master to whom the work is attributed; has cleared up another stretch of that vista at the end of which is Titian at his easel — the fourth of those several harmonies, which, as successively presented, warrant us in looking faithward with ever-growing kindliness of aspect.

Another striking feature of Nature is its *power*.

No contemptible degree of force resides in the muscles of some men — the Samsons and Milos of their time. Huge rocks are lifted, tough oaks are riven, great structures are shaken down by their hands. Many brute animals display still greater muscular strength; witness the elephant, and those gigantic mammals which towered and ruled over the post-tertiary savannas. A combination of animal forces with what are called the mechanical powers often generates measures of force more striking still; and when men stand by such piles as the Egyptian pyramids, they are deeply impressed with the prodigious uplift that must have put those mighty blocks in their high places. But it is to inanimate nature that we must go for our most brilliant examples of physical force. What power in the wind, when, as a tornado, it sweeps along at more than one hundred miles an hour; demolishing mansions, uprooting forests, and lifting ponderous ships far inland on their eddies! What power in the ocean-swell as it tosses an entire navy to the skies with apparently as much ease as if it were a single cockle-shell! — What is this that comes rushing through the landscape with smoky breath and thunderous step, dragging thousands of tons at the pace of winds? Within that flying iron crater is imprisoned one of nature's brawniest forces, steam — throwing off feats of toil with its vaporous arms, which arms of flesh and blood have never even been fabled to do. — What have we here? A few barrels filled with very simple black grains. One has but to drop a spark among them to witness a sudden development of power that shall deafen earth and heaven with its voice, and lift a city into mid-air. — Would you see a mightier energy still? It is the year 1755. An unwonted trembling stirs the air and ground of Lisbon. In a few moments the broad city is in heaps. The plain around runs in waves, like the sea when lashed by a tempest. See — the distant mountain-ranges themselves impetuously shake and rend and topple; Europe, to the Highlands of Scotland, heaves; heaves Africa; heaves the whole broad Atlantic, with all its huge gravities, from the Pillars of Hercules to the New World! When oceans and continents are so tossed and shot aloft, what stalwart shoulders of gas and steam and fire are heaving at the mighty burden! Other forces among us are not small; but this of the earthquake is easy king over all these terrestrial children of pride. Terrestrial, I say: but there are forces not terrestrial which are of a still huger and loftier pattern — celestial forces, to which those of our earth are what the bubble-globules of the children are to the globed worlds of space. When such a planet as Jupiter is moving at the rate of some thirty thousand miles an hour; when such a sun as ours is moving at the rate of some three thousand miles a minute; when such a nebula as our Milky Way, with its eighteen millions of suns, goes wheeling at the same average speed about its center of gravity — there is a momentum for you, a magazine of force by the side of which earthquakes are puny, and all the stormy winds that ever blustered and fought in their fabled caves mere zeros! Some say that there is but one force in all nature — none perhaps more apt to say it than the rejecters of the supernatural — that the forces which pump and assimilate and reject in every blade of grass and leaf and animal fiber; the forces that throb in every ray of light and heat and electricity and magnetism, the forces that swell and toil in every atom of matter, the mechanical forces, the chemical forces, the spiritual forces, the forces here and the forces yonder to the universe's last suburb — that all these forces, with their incomprehensible sum-total of simultaneous impulses, are, after all, but branches of one great central force pushing outward in an infinite variety of directions and forms. If this is so — and who is competent to positively deny it — what a single force that is which can diffuse itself over so immense an area, and divide itself so infinitely, and yet thunder away at special points with such marvelous and terrible energy! If this is not so, still what a wondrous hive of swarming and independent dynamics in this wide nature of ours!

Of course, no one could have the hardihood to say that a nature stocked with such energies as these makes positively *against* the doctrine of a Creator who is himself an Almighty Force. On the contrary, there is a friendly *harmony* between the doctrine and the fact. Were we to find in actual existence a Personal Power to whom nothing is impossible, and learn that he is about to produce a universe, we should *expect* to see produced just such a wonderfully strong nature as we actually have — a nature peopled with strengths, momenta, brawny agencies of most imposing forms and magnitudes. A weak system, a system that is puny in its operations and trifling in its effects, would misrepresent him — shall I not say, would be unworthy of him? Most persons would certainly call it unsuitable; would say that his very nature as an Infinite Power would *demand* of him that he should produce a system that would be continually turning out the greatest results, and so must include forces of the greatest efficiency. When, then, I am *told* that a Sublime Force, who has Almighty for his name, is the author of nature; and I then proceed to place myself out under the open dome amid the pulsings and tossings of innumerable and sometimes immeasurable momenta, and so lay myself honestly open to all their subtle hints and magnetisms; I feel myself silently drinking in predispositions to faith as the exposed fleece drinks in the dew — I feel that the doctrine matches facts; that the asserted creator and creation fit each other as do the die and the face of the coin which it has stamped; that the theory has at least the benediction of yet another verisimilitude and presumption; that Nature, instead of making oath with serene brow and uplifted hand, that there is no wondrous God, significantly asks, with abashed voice, "Is there *not* such a Being?" — in fine, I feel that, as the veil continues to rise from the face of the painting, it reveals still another characteristic of the Great Master, clears up another stretch of that vista which conducts the sight toward Titian bending over his canvas — the fifth of those several harmonies which, as successively presented, warrant us in looking faithward with ever-growing kindliness of aspect.

Another feature of Nature is its remarkable *relation to law.*

Notice *law and its exceptions* — the general steadfastness of modes of being and action in nature, and the occasional breaches in that steadfastness.

On the earth's surface, in its dark interior, in the air and vault above, in the instant present and the ancient past — everywhere, law waves its mighty scepter. Atoms and masses, the ponderables and inponderables, the organic and inorganic, the living and dead — all are evidently subjected in their modes of being and action to certain fixed rules, sometimes particular, but more often covering whole classes of objects. Not a particle floats at random or as a unit: not a leaf grows or falls save according to rigid general principles of science. All chemical elements have their modes and measures of combination to which they steadfastly adhere. All heat, electricity, magnetism, gravity, act according to abiding methods which philosophers have gradually discovered and arranged into the sciences of natural philosophy. The great processes of vegetable and animal life proceed after the same forms and steps, from age to age. The stone beds of the world are formed and modified in certain set ways which are the same now as in the periods anterior to man. Even the weather, so often called fickle, has its stable methods; almost every year bringing to light some new general fact in meteorology, or extending the application of an old one. Day and night succeed each other, every twenty-four hours, without variation. The seasons do not change their order or general character. All of Kepler's and Newton's laws are as operative today as they ever have been since their discovery. The planets shoot round the sun and are circled by their own moons, on substantially the same elliptical orbits, in the same times, and with the same principles of alternate retardation and acceleration as of old. All known changes in the planetary orbits have been found to be bound in a law of periodicity which is apparently invariable. So beyond the solar system. Law still; nothing but law; law everywhere on ten thousand blazing thrones; largely the same laws that prevail in our own system! As far as we can observe — and it is no little that has been observed — those distant orbs reverence the various principles of gravitation and mechanics, and keep as rigidly to their behests, as when the earliest astronomy gazed at them from its rude Uraniberg of a hill-top. And every man of science is well persuaded that, could his observation alight on particular orbs of those remote and twinkling hosts, he would find their minutest details bound up in the chains of the same adamantine regularity that rules our own globe.

So in general we speak. But we must not be understood to speak with absolute precision of language. In this wide scene of steadfast arrangements, there are outbreaks of anomaly — ruptures and rents and dislocations in the habits and ongoings of nature, like those in the strata of the earth. It is a settled law of nature that like shall produce like; yet from perfect animals and vegetables occur occasional monstrosities of organization. It is a settled course of nature that certain substances, called poisons, if freely introduced into animal systems, destroy life; yet now and then a man is found who is even nourished by these agents of destruction. It is a fixed mode of nature that frost withers flat foliage; yet the flat leaves of the wild laurel flourish out our hardest winters. It is a fixed way of nature that the heavenly bodies move in ellipses; yet there is reason to believe that some comets have been found moving on the curve called a parabola. The steadfast habit of nature is against a general planetary deluge, or conflagration, or glacier-period, or destructive convulsion; yet such disasters, if geology may be trusted, have several times occurred, at immense intervals, in the history of our own planet. Great exceptional events; phenomena without fellows through an astonishing stretch of ages; what have the appearance of broad fractures and dislocations of nature, though in reality they may be the rare resultants and accumulations of innumerable natural forces and laws crossing each other in all directions; the entire destruction and rehabilitation of animal and vegetable species — such events have taken place on this globe again and again. Repeatedly has the earth been drowned and torn in pieces. It has been piled with snow and ice from pole to pole. It has been all ablaze and fused. And is it not on the idea of such a conflagration that we can best account for the new stars that have sometimes flashed suddenly on the sight with all the splendor of Venus at its brightest, and, after a few months of changing color and gradual decay, finally disappeared? Thus in the bosom of a general steadfastness are found occasional outbreaks of anomaly. It is as among the geologic strata — where are found faults, dislocations, fissures, and even reversions of those great rock-beds which in general are laid down on a plan of utmost regularity. The course of nature is like some great thoroughfare, which advances through great distances without the slightest solution of its continuity, but at last finds a great river thrust squarely across its track. On this side the thoroughfare, on that side the thoroughfare, and here the broad, deep flow of the bridgeless river — a river worth to the public, it may be, many times what the perfect continuity of the road would be.

Now this much is certain. No one can say that this characteristic of nature makes positively *against* such a steadfast and yet miracle-working God as is affirmed in the Christian Scriptures. Instead of opposition, there is positive *harmony* between the fact and the doctrine. Indeed, such a nature as is observed is just what one would have *expected* to come from such a Creator as is taught. Nay, as general laws are necessary to make science possible, to enable men to forecast and profit by experience, to serve as a basis for all comprehensive business and for all civil government — as the broader and profounder the intelligence, the more it is pleased with and tends to work by general principles, we may say that the very nature and circumstances of Deity would *demand* of him, in case he should create, to create a generally steadfast, law-abiding universe. At the same time, a miracle-worker — one who sees a certain essential imperfection and intractability in second causes, preventing their matching on all occasions the perfection of his ideas; who, moreover, sees it undesirable to allow mere nature to hide its Maker altogether behind its swarming screen, and give to the ideas of necessity and fatality full sweep in human minds — I say, such a being would be under a loud call to provide in the constitution and course of nature such suggestions and prophecies of miracles as would gradually, though perhaps unconsciously to them, prepare the minds of men for those crowning abnormals of the system. He must have the glory of his personal agency glimmer through occasional rents in the uniformity of nature. An anomaly-sprinkled, miracle-suggesting, as well as stable, universe must proceed from his wondrous hand. He would be in conflict with himself were he to produce any other. And when I am *told* of one who is actually just this sort of divinity — both law and miracle: both giver and keeper to an almost infinite extent of moral laws which shall not pass away; while his iron will, throned as supremely in the realm of matter as of morals, yet launches forth into special providences and miracles on extraordinary occasions — when I am told of him, and I then place myself out under the open dome amid the massive but occasionally rifted uniformities, and open myself freely to all their subtle hints and magnetisms, I feel myself softly drinking in predispositions to faith, as the exposed fleece drinks in the dew. I feel that the doctrine and the facts are at one; that the asserted Creator and the observed creation fit each other as do the signet and the seal just stamped; that another verisimilitude spreads blessing, if tremulous, hand over the theory; that Nature instead of sonorously swearing that there is no Divine Being whose double name is Law and Miracle, significantly asks, with abashed and startled tones, "Is there *not* such a Being?" In fine, I feel that, as the veil continues to rise from the face of the painting, it reveals still another characteristic of the Great Master, clears up another stretch of that vista which conducts the sight toward Titian bending over his canvas—the sixth of those several harmonies which, as successively presented, warrant us in looking faithward with ever-growing kindliness of aspect.

Another feature of Nature is its wonderful *relation to time and motion.*

How long has our race existed? The infidel may choose to say a hundred thousand years; none will say less than six thousand. How long has the earth itself existed? The atheist may choose to say, Forever. The geologist, thinking of his coal beds and deltas and rocky strata sown with the bones of extinct species, and of the time requisite for their formation, is sure of several hundred thousand years. How long are the earth and its confederates in the solar system calculated to endure? Geometry declares that no element of decay within endangers the stability of the system of the world. That year which circumscribes our seasons is only three hundred and sixty-five days; but the earth has another year to which this is a mere point — its pole goes nodding through space in a circle which it takes twenty-five thousand years to traverse. What think you of a planet whose winter is more than forty of our years, of a comet whose year is more than thirty of our centuries, of a sun whose year is more than eighteen thousand of our millenniums? All the planetary orbits pass through cycles of changes varying in length from a few centuries to nine thousand, to seventy thousand, to even many million years; but the greatest of these planetary cycles are as nothing compared with those enormous periods which bound the perturbations and express the secular equations of the sun and fixed stars — periods including more years than imagination has ever succeeded in realizing to itself. What amazing longevities! What portentous numerals! They are hieroglyphics of the everlasting. They lift us among the dizziest peaks of the sublime.

These immense periods, interspersed with others exceedingly small, sometimes express an exceedingly slow movement among the powers of nature.

In other cases, the movement with which they are connected is exceedingly rapid. The times consumed in the formation of the coal-beds and rock-strata, and in the long perturbations of the planetary and stellar orbits, are examples of the first class of periods; the years of the planets and stars in their orbits are examples of the second. In the first class, natural forces creep along to their objects with miraculous slowness; in the other, they flash along with swiftness equally astounding. Some orbits gradually lengthen themselves, say an inch in a thousand years. Some of the stars dart along their year of one hundred and eighty thousand centuries at the incomprehensible rate of one hundred and eighty thousand miles an hour. Could we plant ourselves immovably at a certain point in the celestial spaces, and see our sun go sailing by with all its glorious squadrons of planets and moons — sailing down the abyss as if driven by ten thousand hurricanes — would not the sight of such celerity almost irrecoverably daze both senses and spirit?

If, now, one should start up to say that these great cycles, imbosoming unutterable extremes of movement, makes positively *against* an Eternal God who is able to move to his purpose like the light or at a rate so trifling as to be quite imperceptible by human senses, we should laugh his logic to scorn. We know better. These are facts that palpably *agree* with such a theism. Instead of contradicting it, they express a state of things that might have been *expected* from a being who has both unlimited time and unlimited speed at his disposal — who, if he chooses to wait, has never occasion to haste; or, if he chooses to haste, has never occasion to wait — who is alike able to dart on his purpose as if infinite whirlwinds were in his wings, or to approach it at a rate so minute that no human sense can discern the movement in the lapse of generations. Suppose such a God to be about to create a nature, could you not confidently predict after this manner — "This Being of mighty periods will establish mighty periods: this Being who can readily proceed on his endlessly varied designs, at all imaginable and unimaginable rates of speed, will diversify his works with all the velocities." A God who himself has no duration to speak of — if there may be such a God — would never have stored his nature with such mighty cycles; a God who himself never did a swift thing would never have set his laws to spurring on planets and suns so astoundingly; a God who himself never did a slow thing would never have yoked such slow-footed forces to events, as we observe actually dragging at some of them. It is only a God who has substantial forevers on his hands, and who on occasion can lighten and on occasion can linger ineffably along the highway of his purposes, who is properly represented by such a nature. In case he gives any nature at all, his character demands of him to give just this — one expressing his own attributes. So when I am *told* of one whose longevity is eternity, whose orbit of existence has an infinite axis, who reaches an Atonement after slowly beating toward it for forty centuries, who is ages and dispensations in establishing his kingdom in the world, who commonly approaches the punishment of sinners with steps lingering through numberless delays and forbearances, and who yet sometimes yokes steeds of wind and fire and foam to his car — as when some Korah and his company go down quick into the pit; or some Uzziah, profanely grasping an ark, falls dead; or some Ananias and Sapphira, lying to the Holy Ghost, are rushed to judgment in an instant's brief space — when I am told of such a God creating nature; and I then betake myself abroad under the open dome amid those swarming and wondrous orbits of time, now scarred and smoking with the hot hoofs of electric forces, and now pressed by the velvety and trackless feet of forces born of the snail; and frankly lay myself open to all their subtle hints and magnetisms — I feel myself silently drinking in faith, as the exposed fleece drinks in the dew — I feel that there is a significant matching of what we are taught with what we observe; that such theism is on most excellent and embracing terms with Nature, which, so far from saying with uplifted, oath-making hand, 'that there is no Eternal God who, as an agent, is equally at home in an instant and an age,' at least stands tremulously querying, "Is there *not* such a Being?"— in fine, I feel that, as the veil continues to rise from the face of the painting, it reveals still another characteristic of the Great Master, clears up another stretch of that vista which conducts the sight on Titian painting away sublimely at his glowing and glorified landscape — the seventh of those several harmonies which, as successively presented, warrant us in looking faithward with ever-growing kindliness of aspect.

Another feature — the *mysteriousness* of Nature.

Who does not know it? — terrestrial nature is one huge sphinx. She vomits enigmas on us in seas. Riddles too profound for the highest science yet in our possession lurk in every ray of light, in every blade of grass, in every rudest stone. Only some of the coarser facts in relatian to a few things here and there, have been picked up and systematized; and these are what compose our boasted sciences. From surface to center, the earth is choked with mysteries whose stony rind has never yet received a blow, much less a fracture, from the mallet of investigation. Come now, ye great Computers, compute for us how long it will be before the science, which loses itself at the very threshold of the complexities of this world, will be able to swoop down with triumphant wing upon the surfaces and to the fiery centers of those fellow planets that mysteriously weave and interweave paths across the concave, and thoroughly solve the problem of all their swarming contents! A disorderly maze are the apparent paths of the members of our solar system! But you say that the real paths are not as intricate as the apparent. Take your stand, then, at the sun, and observe planets and comets going and coming at all distances and rates of velocity and directions; while around most of the larger planets are similarly moving, other systems of satellites — is it not an intricate as well as a brave sight? Can you see through the mazy plan? But you say that it has been seen through, and planetariums have been made that clearly represent the whole thing to us within a few feet of space. How many centuries and philosophers, O Copernicus — Copernicus, I say, away yonder in the depths of four hundred years ago — did it take to make that orrery and solve that riddle of the system of the world? Indeed, it is yet very far from solution. Astronomers can only completely account for the movements of a system of two bodies. A system of three is quite beyond them; one of a hundred and more bodies, like our solar system, immeasurably beyond them. There is not even a hope that science, with all its dynamical calculuses, will ever overtake this higher problem. But there is a higher problem still. Solar system revolves around solar system; a group of such systems around a similar group; a cluster of such groups around a similar cluster; a firmament of such clusters around a similar firmament. Indeed, as we have seen, the whole universe of stars, with all the countless fleets of planets and moons which they represent, must, according to the law of gravity, revolve about a last center of centers. Let us go to it. Standing at this Heaven — for is not this the dazzling metropolis where dwells the sublime Cesar of the creation — standing at this wondrous point, and looking forth on the countless nebulæ coming and going at all imaginable distances, speeds, and directions — lo, what a glorious scene of bewilderment and unsearchable complexity! It fairly takes away our breath to look. There is no more spirit left in us. If a system of three bodies is too much for the most subtle and comprehensive science yet known, what can ever be done by all coming generations and geniuses, however imperial, toward mastering such labyrinthian immensity of involved orbs?

Now hearken to the Christian Scriptures — affirming a Maker of nature who is himself the mightiest of all enigmas. "Verily, thou art a God that hidest thyself — Canst thou by searching find out God; canst thou find out the Almighty to perfection — It is high as heaven; what canst thou do: deep as hell; what canst thou know?" Does the aspect of nature contradict this doctrine? Who will presume to deny that the incomprehensible materialism about us, to say nothing of the more incomprehensible spiritualism within us, is just what one would expect to find issuing from the hands of an incomprehensible Creator — a being mysteriously without a beginning, mysteriously self-existent, mysteriously able to make the greatest and noblest things out of nothing by simple volition, mysteriously all-knowing, mysteriously unfettered in the application of his power and knowledge by all conditions of space and duration and personal presence, mysteriously Three in One — in short, a being enveloped in a terrible pomp and majesty of sunset-clouds, whose broken lines never permit the orb that glorifies them to appear, even for a moment, in clear and golden contour on our rapt sight. Such a being, setting out to create, would be *likely* to give us the present enigmatic universe, nay — for why state the matter so feebly — would be *sure* to give it. Like every other copious author, he would reproduce his own traits. An unutterable sphinx himself, his creatures would be sphinxes. A nature from the hands of God that I can comprehend, or make any approach to comprehending — preposterous! A creation that to me, with my low place and filmy vision and narrow orbit, is not steeped in seas of mystery — preposterous! If a Jehovah build the temple of nature at all, he will found it on mysteries, frame it with mysteries, cover and dome it with mysteries, pillar and ballast it with mysteries, pave and ceil it with a mosaic of mysteries — surely he will. And when I am *told* of a being whose own nature is an overwhelming problem; whose attributes have no horizon, no zenith, and no nadir; whose ends respect all possible objects and interests, and spread themselves out in plans of boundless vastness whose merest corners and differentials only are visible to men of the widest scope: when I am told of him, and I then place myself out under nature's open dome, amid its Protean inscrutableness of leaf and star, of whole crowded earth and circumventing heavens — the peopled heavens where sweep in inextricable maze the hurricane hosts of advancing and retreating orbs; and open my soul candidly to all their silent suggestions and magnetisms — I feel myself drinking in faith, as the fleece spread out under the stars drinks in the dew — I feel that the facts give embracing arms to the doctrine; that the actual universe, instead of swearing with decisive voice and hand uplift to heaven that there is no inscrutable God, significantly asks with panting whisper and color that comes and goes, "Is there *not* such a Being?" In fine, I feel that our continued lifting of the veil from the painting has disclosed another characteristic of the great master to whom the work is attributed; has cleared up another stretch of that vista which conducts the sight to Titian in the act of glorifying his canvas into the Milanese Coronation-Christ — another of those many harmonies which, as successively presented, warrant us in looking faithward with ever-growing kindliness of aspect.

Such are the facts. I do not say, "Ex uno disce omnes" — as does the naturalist sometimes when he finds a bone. But I say, "Ex multis et maximis disce omnes" — as does a Cuvier when he enthusiastically discovers nearly a complete skeleton. The vestiges we have been viewing are no scant minims. They are no few, narrow, disconnected particulars. On the contrary, they are great genera — the controlling outlines of the picture, the massive frame of the edifice, the sovereign characteristics, the *leading facts and courses of Nature.* As such they give us, so to speak, an extensive taste of Nature. They show us the grand whole, with what is a cardinal, and seems to be an essential, flavor. But we are entitled to assume the self-consistency of Nature, especially in regard to leading features. None so forward to insist on this self-consistency as the modern opponents of our natural theology. Indeed, we only do what is the habit, and the unrebuked habit, and, as we well know, the wondrously successful habit of all Baconian philosophers, when we boldly proceed on the ground that Nature is one.

and that according to what has been discovered of her features is what remains to be discovered. So we are allowed to make broad our conclusions. So we are allowed to say of the integer what we have said of the fraction. Instead of contenting ourselves with affirming that her wisdom, and her vastness, and her power, and many another trait sympathizes with the doctrine of a God, we may go on to say that Nature herself sympathizes with the doctrine. She smiles upon it. She smiles, not as a multum, nor as a majority, but as a total. The whole picture is Titianic. The whole Cosmos is just as if made by God. We might go on lifting the curtain from before her face ten, twenty, never so many times, and always with the same result. Never a break in the verdict. Never a rise of the veil that says, "Lo, here the facts are out of sympathy with the doctrine." Never a trait of that queenly face comes drifting into view to say, "Lo, here is something that looks as if there were no God." For now some thousands of years our natural knowledge has been advancing, and the envious curtain has been rising, step by step; and never yet, I am bold to say, has the observer, after carefully looking on the picture without and carefully listening to the voice within, ever heard any other words than these: "Just as if made by God — Just as if made by God!" Should we go on lifting the curtain till it is looped up to the very ceiling of the uttermost heaven, we should find all things in harmony with that sonorous verdict that has already come surging in upon us from the four cardinals, like so many trade-winds: "Just as if—Just as if." Oh be sure we may go on from speaking of the attitude of parts of Nature to speaking of the attitude of Nature herself! She, this modern goddess, is no enraged Bellona, shaking her spear in the face of the doctrine. She is not even adverse after the softest and sweetest-tempered Cyprian fashion. Her ways are most kindly and cordial. She embraces the doctrine. As we see ourselves pictured in the glad, beaming eyes of the long-lost friend whom we hold in our arms, so the Theism sees itself pictured in the cordial eyes of embracing Nature. Their voices chord. They are phone and antiphone. They are parts in the same rich chorus. The doctrine is merely the shadow which Nature casts on a book. And I think I am modest in my asking when I ask that so many and great verisimilitudes be considered as completely clearing the ground for faith, and as standing at the open gate of the judgment with bright and welcoming faces, ready to grant possession at the first summons of the positive evidence. I think I might reasonably ask much more. I might lay stress on a great difference between our case and that of the supposed painting by Titian; might point you to the fact that whatever traits of that master are found in that picture, are obviously such in nature and degree as lie fully within the power of many a human nature; whereas many of those traits of Nature which we have just come from viewing are presumptively and enormously impossible to any agency short of a Divine. But let this be waived. Let me only claim that with every new harmony which the rising curtain has discovered, my mind has rationally moved faithward: and that now, when these harmonies have been found many and potential enough to give character to the whole magnificent Outspread, my fleece, exposed through the long night till full break of day, is rationally wringing wet with the dew of predispositions to faith; and that, at least just as soon as the positive evidence pushes its orb above the horizon, I may hold it fair and scientific to allow each tiny drop to be transfigured from the silver of a predisposition to faith into the gold of faith itself — making, as I think, the true golden fleece of Colchis for modern times, for which all Argos should sail and all heroes strive.

VII.

NEED OF GOD.

Τὰ Θεμέλια τοῦ οὐρανοῦ ἐσπαράχθησαν.

Πάντα δε Δὶος κέχρημεθα πάντες. — *Aratus.*

VII. Need of God.

SEVENTH LECTURE.

NEED OF GOD.

WE are not only beckoned toward Theism by the harmonies between it and Nature, but we are beckoned toward it even more strongly by that crying Need of God which a little examination will discover.

It is a very striking thing — the general wish for a God that exists among virtuous men. We seem to see that, from the beginning till now, there never has been a person who has honestly tried to follow his conscience who has not honestly desired to find Him a reality. And, the more conscientious and exemplary the man has grown, the more unwilling has he become to part with the idea of such a Being with His universal and glorious providence. There are persons who would as soon be without God as not; nay, there are those who could hardly hear pleasanter news than that the whole Theistic argument has been fairly overturned from the foundation, and the impossibility of a God proved beyond all possibility of denial. Oh, how scoffing

Voltaire and licentious Rousseau and bloated Paine would have clapped their hands and shouted, could they have fallen on some wonderful geometry which by its rigid demonstrations could compel even the most unwilling to give up their last plea for Deity! Oh, how the high-handed evil-doers of every name, sinning and impetuously bent on sinning, would congratulate themselves, could it be made as plain as day that such a machine as a thinking, embodied man was created by chance; that chance fitted up the earth as his convenient home, and hung out the heavens above him with the blazon of stars and suns! But as soon as a sinner has made up his mind conclusively against sin, and has fully committed himself for endless war upon it in all its forms, then he ceases to be averse or indifferent to the Divine existence; then he begins to positively like the idea, as including that of a righteous Divine government; then he begins to cling to it, to bless it, to feel that he cannot do without it; and as he goes on to higher and higher grades of virtue, the feeling in behalf of God gradually deepens into a profound hunger and thirst. He says, "My soul thirsts for God, for the living God." To him an offer to disprove God would be an offer to make the universe an awful solitude and desert.

For, he looks around and sees all things wavering and changing like the baseless fabric of a vision; and even those things which seemed most solid and stable, and to which multitudes had ventured to anchor themselves as to so much eternity, quietly or violently slip away and dissolve and leave not a rack behind. And is there nothing stable in this shifting scene? Is there no rock that lifts its head high above the fluctuations, and shakes not, trembles not, for aye, though all things rock and break and die around? Is there no one point of stability around which poor man may gather his affections and trusts and reverences and hopes without being liable every moment to see his center vanish into thin air? At such a moment it is a great comfort for him to look up and see, or think he sees, a God of infinite goodness standing fast forever in the wide sea of change — a still, green continent amid the tossing ocean, under the lee of which the ship may pass the night without fear of finding its shelter gone in the morning, like a bank of clouds — a pole-star that is always there, though every other orb at least rises and sets. Here at last is repose — here at length something to build upon. Here is a friend who will never die — here a glorious providence that will never have done reigning — here an unspeakable lover who will never grow cold — here the perpetual overruler of his mistakes and sins; enlightener of his conscience; eradicator of his depravity; educator of his incompetency; support and consolation under his trials, come when and where they may. For, in the time of distress, he can look away to the righteous throne of the Infinite Disposer, and see stars seeding his darkness. And when the hour is midnight, he is not unused to say, "What should I do now without a God to go to?" He goes to God, and his heart is bound up. And it is a gladness to feel that he can do this forever, could his sorrow last so long; for his heart sings amid its groans, "O Lord, Thy years are throughout all generations." In fine, it is such a comfort to have a God, that it would be day turned to night were his Theism to become an atheism.

Now truth is kindred and polar to goodness; and so the desires of good men are most apt to harmonize with and point at the truth. The most virtuous have most affinity with the truth; are most free from prejudices and intellectual obliquities; are the most fair-minded, earnest, and laborious in their inquiries; and so their opinions and tendencies to opinions are most apt to be correct. — Also, the mind when virtuous is in its soundest and most normal state; and the features which belong to the whole class of virtuous minds, in proportion to their virtue, are natural to the virtuous mind. But, on observation, we find that there are, outside of this case and two or three other mooted cases — such as that of the desire for immortality — no desires natural to a sound human constitution for objects which do not exist. Thus a desire for food, for friendship, for knowledge, for reputation, for society, for liberty, for freedom from pain, is natural to every sound human constitution; and the object of each of these desires actually exists. The food exists to meet the hunger; the beauty exists to meet the taste for beauty; there is knowledge, society, health, to meet the natural relish for each of these things. Indeed, you cannot point out a single desire natural to a sound human nature to which there is not an answering object somewhere; but, on the contrary, such answering object is certainly known to exist, outside of the two or three disputed cases, like that under consideration. Hence we must conclude that the desire for God, which is natural to virtuous or normal minds, has over against it in the outward universe such a God to fulfill it. — But what I am more particularly concerned with at present is the strong testimony which the attitude of virtuous men toward Theism gives to the need of God. The utmost we can be asked to allow, in regard to the disposition of such men to believe in God, is that it comes from a desire for such a Being. They wish a God, and this feeling naturally draws faith after it. The wish is father to the thought. Supposing this to be so — supposing that the whole case is reduced to one of desire for a God — how comes such a desire to exist, to exist in proportion to virtue, to exist in the most virtuous as an intense craving? Two answers can be given. One is that such craving is the instinctive aching of a great vacancy for a great supply: the other is that the craving comes from an intellectual judgment that a God is vastly desirable.

In both cases we have human nature in its soundest state testifying loudly to the need of God; in the one case by its natural instincts, and in the other by its intellectual convictions. And the need testified to is organic, because it belongs to essential human nature in its most normal state — is generic, because it belongs to a great class of beings — is most important in kind, because it belongs to the most important of earthly beings in their most important relations. Whoever has virtue is Agamemnon, king of men. When you see a fairly hung vane straining away at the west as if a gale were blowing from that quarter, make sure there sits the wind. And when you see our best and truest nature pointing Godward with a snowy finger as determined and intense as was ever chiseled out of marble, make sure that a broad organic need of God is invoking that motionless digit.

But the traveler may not only wish to be reliably pointed in the direction of a great city; he may wish to advance, and see it with his own eyes. It is thus his impression of it will become more correct, vivid, and enduring. And we shall have that Need of God which the polarities of good men point at, directly under our eye, if we go on to consider the *practical influence* of Theism.

It has been usual for leading unbelievers to confess the excellent moral tendency of the doctrine of a righteous Divine Ruler. And ask any man of ordinary sense and observation, putting him on his honor and conscience to speak frankly — ask him whether he does not really think that a solid faith in such a Being would, on the whole, be a greatly better thing for his son and all connected with him than disbelief or unbelief would be — what would be the answer? He might not speak it, but ere a moment could elapse he would think it. "Practically," would he say to himself, "it is better that my child should believe. Whatever may be the abstract truth in the case, I cannot deny that such a belief is likely to be followed by better results to himself and to all within his sphere of influence than the absence of that belief." And could the ideas — perhaps exceedingly vague, fragmental, and disordered — which form his reasons for this view be formally drawn out and expressed, they might be found stated somewhat as follows: — "Character is the great interest of a man. An unprincipled and wicked life is low, disgraceful, and destructive. And it does not admit of reasonable question that those believing in a righteous Divine Ruler — that is, all who believe in God at all — are under stronger moral restraints than others; that the more profoundly society is impressed with the conviction of an Infinite One above it who loves righteousness and hates iniquity, and will without fail hold men to account for their conduct and the secret evil of their hearts, the more orderly and virtuous it will be. The bare presence to the mind of such an idea — of an idea so majestic and pure, so grand and lofty and thrilling, as that of a God actually living and reigning, never beginning and never ending, knowing virtually all things and powerful to do virtually all things, true and just and good in the highest conceivable degree, having for His sweet name Unfathomable and Shoreless Love — must tend powerfully to educate the moral sense; to expand, elevate, and purify the soul. It can be nothing less than one of the greatest of moral cultivators. And, besides its power as a great and pure idea, the conception of God as an actual existence must have vast power to restrain from evil and encourage to good by the strong appeals it makes to the principles of hope and fear. A sinner cannot steadily look at the thought of a just God without trembling; and even a shadowy impression of such a Being leaves a voice in the heart which says, 'Be warned: if you are wise, you will cease to do evil.' A good man cannot hold steadily before him the thought of an Infinite Being taking account of every right act and rejoicing over it, without brightly hoping: and even a vague, embryonic impression of such a Being leaves words in the heart which say, 'Blessed be thou of the Lord; go on and prosper.' The one is restrained, and the other is encouraged — greatly and necessarily. Without such restraint and encouragement, the water-line of morals in this world would be far below its present level. Why, consider what a God this is who men say reigns in glory and righteousness everlasting; and is he tc whom this mighty Personality is a solemn reality under no greater pressure to virtue than he to whom such a Being is a fable or an uncertainty? Sure we are that there are few thoughtful men who would be so unreasonable as to think it. As well almost might they think that objects on the surface of the earth are just as likely to fly off from it with as without the gigantic forces of gravitation steadily drawing from all points toward the center. No: from every point of view the natural tendency of faith in such a God is toward virtue, toward virtue only, toward virtue vastly. And though it is true that moral beings, from their very nature, are competent to so resist this natural tendency as to make it the source of increased guilt and misery, and often do so, yet in the majority of cases it will not be done. The more faith, the more motive against sin; the more motive against sin a man has, the more likely he is to restrain it; and, if the individual is more likely to restrain it, the community at large will actually restrain it better. Men universally act on the principle that if they can make persons believe vividly that it is their interest to take a certain course, the effect in that direction will be favorable in a majority of cases. These rational deductions are confirmed by observation. Comparing together large communities, one observes that those are the most orderly and moral in which faith in a righteous

Divine Governor prevails to the greatest extent. We have on record only one instance of a nation of atheists; and what a frightful state of disorder, demoralization, and terror accompanied the phenomenon, the world, and especially Paris, will not soon forget. Milton's description of Sin is not too strong to suit atheistic France of the Revolution:

> 'Seemed woman to the waist, and fair,
> But ended foul in many a scaly fold
> Voluminous and vast, a serpent arm'd
> With mortal sting: about her middle round
> A cry of hell-hounds never ceasing barked
> With wide Cerberian mouths full loud, and rung
> A hideous peal.'

No civilized people ever gave so bloody and foul a chapter to history. May history never receive such another! Further, it is observable that individuals with habitually very vivid and strong Theistic faith are almost, if not quite, always very virtuous; certainly very much more free from misconduct than other persons. Nearly all positive rejecters of God, and indeed nearly all professed skeptics as to Him, known to the reading public, have been public lepers both as to the principles and practice of common morals — have fought against, not only the doctrine of a God, but also the doctrine of moral distinctions and all the ten commandments, both with their pens and with their lives."

So might soliloquize almost any intelligent father. On the basis of mere broken hints of such facts he might well desire faith for his child. Which the more friendly to virtue, faith or its absence — how can he for a moment be at a loss how to answer the question? There can be no comparison between the two. As motive to virtue one is everything, and the other nothing. On the one hand you have a giant statured like the demigods of fable, clothed in prodigious sinew and brawn, and making the earth and its starry dome to shake at every step — such a giant pushing us toward all open and secret righteousness with might and main. On the other hand you have a pigmy asthmatic skeleton, scarce able to stand or breathe alone, laying on us the feathery touch of a single skinny finger, which, ten to one, is not noticed by us at all, save as a chill or a paralysis on all virtuous endeavor. Sure we are that, instead of giving the smallest pressure in the right direction, it rather goes to chill off the soul as with the damps of the grave from all excellent activities. When we see that for the leaders in unbelief to discard God is generally to cast off righteousness, lose conscience, and unlearn the very theory of obligation — when, in the only example the world has ever seen of organized atheism, we see it shaking society like an earthquake, and crushing alike the sanctities of home, the rights of the citizen, and the authority of law — when we consider the instincts of parents, the confessions of atheistic apostles, and the very nature of the case, it is impossible to give any but the most gloomy account of the practical influence of any system of living which does not positively recognize a Divine government. Strike out the present almost universal idea of such a government, and we are bound to believe that the present level of morals in the world would sink with startling swiftness and prodigious ebb so as to show the very central sands. If some misanthrope would turn the world into such a state that, in comparison with it, the present state should be an Eden-garden, let him find some way to extinguish from the world all belief or suspicion of a Divine existence. This would bring upon us the briers and thorns of the wilderness as nothing else would. If any father wishes to bring up his family to be impetuous and bold for every folly and for every crime, ready to trample on all civil and social and filial duties; sad tempests and plagues always and everywhere — then let him bring them up to think that there is no God above to watch and deal with them according to their works. In a word, Theism is the salt of the world. We should be nothing but a putrefying corpse without it. And, on the other hand, should all men come to believe in God as they believe in the oceans and mountains — as vividly and profoundly — the world would become a beautiful life, and its present faded and hectic cheek would so nobly round and flush with virtue's health that one would hardly recognize it.

Thus faith in God is greatly favorable to virtue. Being so, it is greatly favorable to *happiness* here and to prospects of it hereafter. Let one consult the general admission of mankind; let him consult his own experience and observation; let him consult the very definition of virtue, which includes the idea of acting in harmony with the nature of things; and he must feel that the happiness of a man in this world depends more on his relation to virtue than on all other things put together. Beyond a doubt, goodness is the great sunshine-maker. There is a world of poetry and of truth — and not more of one than of the other — in that antique phrase, Sun of Righteousness. The kind of pleasure virtue gives is incomparably more pure, penetrating, lasting, and elevated than any other. It is sweet as the ambrosia and nectar of the gods. It can make a little heaven in the absence of all things else: all things else leave us but an empty and pricking satisfaction without it. It gives a quiet conscience, governed passions, benevolent affections, concord with natural laws, and generally sublime hopes. While leaving us as fair candidates as others for all forms of worldly good, it doubles our faculty for enjoying them: while leaving us to no more and greater trials than befall others, it provides us with sevenfold consolations under them. Virtue does indeed forbid us certain pleasures. It will not allow us to drink of everything that goes by that much abused name, or that really gives much present gratification — perhaps foamy, passionate, incarnadined Falernian. But the gratifications which virtue refuses are well known to leave behind them so many bitter tastes and pains as to make them, on the whole scheme of life, sorrows. Our interest of earthly happiness ought to thank virtue for cutting us off from such apples of Sodom and grapes of Gomorrah. — And is there another life beyond this? If so, the virtuous man stands by far the best chance in regard to it. Should a Holy Governing God prove real, such a man must be saved and glorified; should He prove but an empty name, such a man can hardly be worse off than his unvirtuous neighbor. Who believes that, Christendom being searched through, a single intelligent man could be found, who, looking merely at the chances for safety and happiness after death, had not rather live and die as a good man than as a bad one? Every person is absolutely certain that his future state would not be prejudiced by a virtuous character and career; and he is not certain but that it would be ruined by the want of these.

Then, as to the bearing of virtue on the *usefulness* of a man. When we say that a certain man is virtuous, we in effect say that he habitually discharges what seem to him his duties in all directions. He stands nobly pledged to his family, to his neighbors, to his countrymen, to his race, and even to brute and inanimate nature. We may be sure of his being, in the main and according to his light, a good son, brother, father, employer, servant, citizen, ruler, subject. Even the beasts and herbs of the field shall be the better and thriftier for him The very nature, as well as the history of virtue assures all this. On the other hand, the unrighteousness which atheism fosters assures nothing of the sort. It leaves the way open for every sort and degree of trespass on the interests of others. A man may plague his friends, and curse his neighborhood, and betray his country — may belch out corruptions and injuries of every name on all around him as stormily and profusely as ever did volcano its destructive ashes and lava — and still be a capital atheist. If he chooses to commit the grossest possible outrage on society, and then assert that he does not believe in God, we cannot say that his assertion and his deed are mutually inconsistent. Go to that man in prison on charge of having murdered his own loving, self-sacrificing, and beseeching mother. Say to him, "Man, what think you about this matter of a God?" "Think!" he shall say, "why I don't believe in Him — there is no such Being." Would such an answer go to strengthen any lurking idea of his innocence which you may entertain? Would you feel like grasping his hand, and exclaiming, "My dear sir, it must be that you have been unjustly accused! You a matricide! You an atheist, and yet do such a crime! You honestly convinced that there is no God to bring the wicked to account, and yet lift murdering hand on your own mother! Incredible! The friends of justice must at once look to your liberation."

How preposterous such language would sound to even the prisoner himself! He would think you dealing in severest irony. He knows, and is sure that everybody knows, that the loss of that faith which he had when a boy has done nothing for him morally, and could do nothing whatever, save to mightily lift the checks from his depravity. He knows, and is sure that everybody knows, that godlessness is just the thing from which to expect all sorts of harmful outgoings on the world around — from men down to the dog that crouches at his feet, and to the sapling that stands fearing the wasteful axe.

For the full scientific significance of these facts we must wait a little. After gathering some further particulars of the same general character, we will proceed to question science in regard to the value of the whole. But I will ask you to just notice in passing that, in case there is a righteous Divine Ruler, this immense utility to great classes of beings of faith in Him is just what we ought to find. He would be likely to make a system to which He should be the necessary complement; a system which should perpetually call for Him; a system which should find its highest repose, satisfaction, and uses of all sorts in a cordial recognition of Him. But, without a God, the fact would not be positively likely to be as we find it: on the contrary, it would be positively improbable. Is it not matter of universal admission that it is in general best for even the individual to avoid error, that generally it is best for him to see things as they are? Much more sure is the doctrine — much more does it square with the convictions, experience, and proceedings of mankind, that for a whole class it is substantially never better to believe, especially permanently, in a falsehood than in the conflicting truth. Indeed, I am confident that not a single instance of the kind can be discovered by any amount of research. And if the class in question be the whole race of men, then indeed it can be concisely proved that such an instance is impossible — there cannot be an instance of universal faith in a falsehood proving vastly more serviceable to the public than faith in the opposed truth. If it is extremely desirable, all things considered, that men at large should hold to a given error, then true benevolence requires you to promote that false belief in yourself and others to the extent of your ability — no matter though at the outset you know it to be unwarranted by facts. This being so, unless there is generic antagonism between the dictates of general benevolence and those of duty, between a useful course of conduct and a righteous one — which is allowed by no credible and tolerable theory of morals — it is your duty to abuse your reason, to make and love a lie, to employ prejudices and sophisms and all sorts of intellectual trickery to impose on yourself and others. Samson must put out his own eyes and those of all Israel besides. Who believes this? Are we prepared to give up the most instinctive, universally received, and fundamental principles of morals? If in any one case it is not merely right, but a positive duty, to practice such moral jugglery, then there is no radical and inherent distinction between right and wrong, and one may carry his principles along with his crops to market without compunction, or run them up for his amusement to the side of some highest vane, to bear it company as it turns easily by a breath to every point of the compass. The doctrine is simply abominable. Get thee hence, Satan! Any campanile would be dishonored by such a weather-cock. One cannot persuade the human mind to accept falsehood as true on good grounds. Just arguments never prove an error. Natural, judicious, and honest processes of thought have no tendency to carry us to wrong conclusions. If one manages to escape the grasp of known truth, it must be by pettifoggings, cheats, treacheries, false swearings against his own reason and conscience — it must be by shameful twistings, turnings, doublings, and even metamorphoses of his better nature. Behold the struggling Proteus! At last the god becomes a swine. Is it his duty to do this? Is he at liberty to do it? How dare he so debase his divinity? How can he help despising his ugly self wallowing in that sty? Avaunt Proteus, Machiavelli, Mephistophiles — we will have none of you! Away with such repulsive and destructive principles — on the point of the longest and most nonconducting of spears! It is as if one had grasped the battery of a torpedo!

But — theory apart — a bad practical influence is a noted characteristic of falsehood. I do not say there may not be seeming exceptions. But I do say, that, in general, it is unfavorable to our interests to believe a falsehood, or to fail to believe the truth. Otherwise, to say the least, we might as well have one opinion as another; mistake would be as likely to be serviceable as just views; and all pains to investigate would be foolish — a thing that nobody believes, and the contrary of which everybody assumes in the affairs of actual life. Witness, ye sciences — sweating away at your observation and experiment and induction! Witness, ye arts and trades — straining away at the toil of inventors and adapters of inventions! Witness, ye sagacious business men of every name — knitting your questioning, forecasting, anxious brows over ledgers, markets, products! What mean ye all — unless it be that knowledge is better than ignorance; that facts are the mine out of which men are to dig their prosperity; that he who best knows things as they are, and best adjusts his conduct to them, has advantage over every competitor? This is what all this circumspect and thoughtful activity of the world means. And means wisely. For this is only saying that the engineer who lays out his road across the continent after copious and careful surveys, and with eyes wide open on the correlations of hill and valley and marsh and river, will surely do better by his company than if he had gone blindfolded to his work; that the sailor who, with alert hand on the wheel, watches closely compass and chart and sea and sky, will be more likely to make a prosperous voyage than if he had chosen to sleep or to be drunken; that a blind traveller who paces at random about Switzerland is more likely to come to harm than if, with eye like a sunbeam, he were carefully noting every step he took among the torrents and glaciers and chasms and terrible precipices. Civilization is better than barbarism. The Nineteenth Century is better than the Age of Bronze. The United States are better than Dahomey. And that truth, secular and religious, out of the acquisition and use of which all this mighty difference has slowly grown, is very profitable truth. So generic utility as a trait of truth, and generic hurtfulness as a trait of error, stand demonstrated on an immense scale in the history of the world. The conduct of the world admits it, the experience of the world proves it, the very foundations of morality demand it. And when we find the Doctrine of God with this bright distinction shining on its breast, like the jeweled star that betokens an emperor; when we find it so radically and vastly useful to the virtue, happiness, and usefulness of the broad class of men, and so to the welfare of all the dependent animal races with their still broader domain — our heads instinctively sink upon our breasts, and we do homage as in the august presence of that unchallenged sovereign whose name is Truth.

I pass to another point. The polarities of good men point toward a great need of God. The practical influence of Theism places that need actually under our eye somewhat in detail. A consideration of the effects of a *direct Divine action* on the universe will manifest the need still more fully and impressively. This will show great Rome from the summit of the Capitoline. The distant guide-boards of the Appian have reliably pointed us toward the city. Coming to the gate, we have caught partial views of the interior. And now, at last arrived at the city-heart and perched far above Rock Tarpeian, we proceed for a moment to take the supreme view. All the monuments are before us. On the one hand is the old city, with its arches and temples and Coliseum of the mighty past; on the other is the new city, with its palaces and basilicas and Vatican of the mightier present. The most impressive sight of all!

Conceive of a direct Divine action on the universe. Hardly anything can be surer than that such action cannot be the slightest harm to the general interest of the system, but must, on the contrary, promote that interest infinitely by force of infinite faculties acting through infinite years. With One standing at the wheel who can at a glance see through the whole system as related to both space and duration; who out of a loving heart wants to make out of that vast empire the best He can, and who can on occasion instantaneously and forever send out to its remotest extremities the help of resources virtually immeasurable; and who is actually counteracting, guiding, propelling everywhere in all that stupendous sovereignty—enlightening ignorance, comforting sorrow, restraining sin, stimulating holiness, forcing brute energies and elements along their appropriate channels by an overmastering omnipotence—with such a Being the system of Nature as a whole is sure to reach, not only the highest destiny in the nature of things possible, but also one infinitely in advance of what would have been realized without Him. Individual interests found in irreconcilable conflict with the general interest will have to suffer; but the general interest itself, in all its huge proportions, will steadily have the benefit of a still huger wisdom and power working endlessly in its behalf. And this will be an immeasurable benefit.

According to the doctrine of chances it is morally certain that without a God the issue of the universe will not be the best possible. Further, without a God there are as many chances of the issue being bad as of its being good. Still further, it is possible that the issue will be calamitous beyond expression. On the ground of no presiding Deity we cannot venture to predict the last or balanced result of anything. What is there whose history we can follow through all the intricate workings and interworkings and counterworkings of an infinite number of independent agencies upon it, and so gather up its general significance, whether fortunate or unfortunate, best possible or worst possible? The web that is being woven is so large, and such myriads of shuttles shoot before our eyes in as many different directions, that we cannot make out the pattern. Here we see a bright thread, and there a dark one: but what figure will come out in the end no mere looking into endless mazes will tell us. But give us the fact that there is an infinitely accomplished Being at the head of the universe, and we can see in a moment what sort of a pattern, with its particolored threads and swarming shuttles, the great loom of existence is weaving. Behold the most perfect picture the case admits of! Behold a tapestry Gobelin whose precious threads spell out the sweetness of celestial landscapes — fit hangings for the palace of the Eternal! The system of things will turn out happily, and as happily as, in the nature of things, it possibly can. It cannot be a curse or a failure, as without a God it may be. It must be a positive blessing, as without a God there is no positive probability of its being. It must be the greatest blessing possible, as without a God it is certain not to be. Nay, this greatest possible blessing must be an infinite one, as without a God it is certain to be at an infinite remove from being. The perfect goodness of God would never have allowed Him to bring into being a system out of which He could not extract more good than hurt: once in being, that system cannot but be gloriously advantaged by the eternal action upon it for good of that glorious magazine of forces and resources of which His name is the magnificent cordon. Such forces cannot fail of effects commensurate with themselves. All forces actually exerted on an object are effectual, and effectual in proportion to their degree. If they fail to produce a positive movement in the direction in which they act, they at least destroy so much force in the opposite direction. Hence the resultant of a God on the well-being of the universe must be unspeakably precious. The difference between what it is with Him and what it would be without Him, is solid infinity. The two are as far apart as are the poles of Nature. Just think of an infinite Being working with undecaying diligence through everlasting years for the best good of His universal creatures—giving to the great enterprise all the wealth of His loving heart, all the resources of His unbounded intelligence, all the energies of His almighty arm! Is it in the power of such minds as ours to compute the value of such an agency as this? Astonished Arithmetic refuses to undertake the problem. She refuses even to seriously look at it. What have her puny multiplication tables to do with such an optimism as this? To be without a God is for the universe to suffer an infinite loss. It needs Him as it needs to escape an infinite evil. Measureless Need! Where is the fathom-line that can sound it—where the ship or the electricity that can log across its broad expanses —where the aeronaut or the telescope even that can shoot upward to the foamy crest of its ground swell?

When you see a little orphan child — weak, sickly, willful, destitute, tempted — you are touched with compassion. How much he needs a father! How much he needs a father to protect him, to counsel him, to govern him, to educate him, to provide for his future in almost every respect! Such is the spontaneous feeling of every kind and thoughtful man as he sees the poor boy wandering about in his rags, able to do little or nothing for himself, full of fears and ignorance, beset on all sides with great dangers and miseries, prone by nature to almost every kind of evil, and already showing the beginnings of many bad habits and disorders in body and mind. How much he needs a father!

So I feel when I look about on the fatherless world which atheism presents to us. Ah, such a world — let no one tell me that it does not need a God! It is weak and sickly; and needs a strong Divine arm to lean upon, a strong Divine tenderness to nurse and shield it. It is a world in rags, cold, hungry, thirsty, wandering about without shelter under inclement skies; it needs a Heavenly Father to care for it, and give it home and fireside and raiment and daily bread. It is a sorrowful world — it needs a Heavenly and Omnipresent Consoler; a world full of temptations and dangers — it needs a Heavenly and Omnipresent Protector; a world full of wrong tendencies and actual disorders, willful and wayward and corrupt to a miracle — it needs a Wise, Omnipotent, and yet Pitiful Heavenly Governor; a world full of ignorance and error — it needs a Heavenly and Omnipotent Counselor and Enlightener.

From what now is in this world, imagine what would be in all worlds were Nature thoroughly drained of a Divine existence and government. It would be a most terrible state of things. My account of it has not been too strong. From the nature of the case it beggars description. An Infinite Force that lays itself out most unsparingly and wisely and eternally for the best good of Nature, could not be subtracted from it without infinite damage. It could not be added to it without infinite gain.

We are now to ask carefully for the scientific import of this Need of God.

It is a canon of modern science that whatever is needed to complement a race of beings in any constitutional respect, always exists somewhere, or is attainable. Take, for example, the race of men. Whatever is needed to match and make fully available any of our natural mechanisms, faculties, aptitudes, traits, so that there may be no waste — so that we may have the full benefit of such powers as normally belong to us — whatever is. needed to do this always exists, or, at least, can be made to exist. Thus men need air. They need it to turn to some account the breathing mechanism of their bodies. Accordingly air exists. — Men need light. They need it to make some use of their eyes. These organs were as good as thrown away without it. Accordingly light exists. — Men need odors and sounds. They need them to turn to some account the organs for smelling and hearing. Accordingly odors and means of sound exist. — Men need food and heat. They need these things to sustain the action of every bodily organ. We might as well not have the organs as to have them frozen and strengthless. They would be sacrificed. Accordingly heat and food exist. — Men need knowledge, friendship, government, virtue. They need these things in order that the constitutional faculties and demands for them which human nature possesses may not be quite useless and worse than useless. We should be wretched with nothing or next to nothing to answer to these hungry and thirsty parts of our nature. Accordingly they do not hunger and thirst in vain. There is knowledge to be acquired; love to be given and taken; government to restrain, direct, and compose the social state; virtue wherewith to exalt and felicitate ourselves. — In short, Nature builds no reservoir which she does not, sooner or later, use; forges no tool which she has never occasion for. She is self-congruous and self-fulfilling.

Not only do such things exist as are needed to prevent an entire waste of any constitutional faculty or trait of mankind, but also all such things as are needed to prevent even a partial waste of it. Nature is thrifty. She is a great utilizer. She abhors waste in all its degrees. She gathers up the fragments that nothing be lost. And so provision is made for the full use of everything that really belongs to our nature. The best condition of each part is made possible. A pendulum may describe a very small arc, or a large one, or a whole semicircle. An eye may see little, or much, or as much as belongs to an eye of the soundest state and wisest culture. It is this last measure of activity and use that Nature provides for. She sees to it that every part of human nature is provided with the means of describing its full semicircle. For example. Men need not only air, light, sound, odors, food, heat, knowledge, friendship, government, virtue — without which certainly constitutional traits would be totally unavailable — but they also need certain measures and varieties of these things in order that those parts of our nature to which they have respect may be available in the fullest degree which their natures allow. Unless the air has a certain density, the lungs begin to labor. If the light has not a certain tone and intensity, the eye is more or less crippled in its performance. Unless the sounds and odors are tempered and varied to a certain extent, the ear and nostril cannot do full service. Unless we have a variety of food and certain limits of temperature, our bodies weaken throughout, and every faculty walks with trembling knees. If we do not have large measures and many sorts of knowledge and friendship and virtue, the most possible is not made of our faculties for these things — we are not suitably fed and equipped for the best experience and service which our powers allow — the oil does not fill the capacious bowl, the wick is too small for its large tube, and accordingly the flame flickers on the high silver socket where it ought to burn steadily, and the photosphere is pale which ought to be flooded with light. Such is the need. Nature has provided accordingly. That density of air, that measure of light, that variety of food, that range of knowledge and friendship and virtue, which best suits our powers, is either actual or attainable. We find that if much change is made in either of these respects we at once begin to suffer. Our vitality abates. The system becomes depressed. Strength ebbs away at every pore, and every organ gives sign of embarrassed action. We discover that our circumstances stand well adjusted to our natures. What the race needs to best utilize its various faculties the race has. The stature of the supply matches the stature of the demand. Nature builds no reservoir twice as large as she can use; nor forges a tool twice as sharp and massive as she has occasion for. She is self-congruous and self-fulfilling.

These are a few examples of the law that whatever is needed to turn to account, and even the fullest account, the traits normal to any class of beings, exists somewhere or is attainable. Of course it is not claimed that this law has been verified by an actual examination of every single case of such need in even a single department of great Nature. But it is claimed that so extensive an examination of particular cases has been made as to put the law on the sure footing of inductive science. We have had an immense experience. And all our experience has been one way. It has been to the effect that Nature, within the field stated, does nothing by halves. She does not stop at fractions of enterprises. She never forsakes a part till it becomes a whole. Her works are often a process; not seldom the process is long; but provision is always made for finishing up in a congruous manner whatever she has undertaken. Many human works are finally forsaken at various stages of incompleteness — schemes, machines, edifices, books. You cannot infer from the unfinished tower of Cologne, or from the unfurnished niches on the minster of Milan, that it ever will or can be supplemented into completeness. Not so with the generic works of Nature. She is no Michael Angelo — leaving piles of unfinished productions. She is no Livy — certain chapters given, and then a hopeless "Cætera desunt." All her parts bid us look for wholes. Each fraction of hers proclaims that its integer is come or coming. Have you found such a fraction? Be sure that all the things needed to round it out into completeness exist somewhere, either *in esse* or *in posse*—as sure as you are when you see the red of the spectrum that its complementary colors are not far distant—as sure as you are when you see a crescent moon that the rest of the sphere is by its side, though for the present unilluminated. Look more closely, and perhaps you will faintly discover the old moon in the arms of the new. Look more closely, and perhaps you will discover over against yonder organic need in Nature that full supply of the need which Nature has provided. But whether you discover it or not, make sure that the supply exists or is attainable. Nature does not waste herself. She has no fondness for throwing herself away, either wholly or in part. If you find one of her reservoirs, make sure that there is something to put in it, and as much as it will hold; if you find one of her tools, make sure that it has something to do, and as much as it can do well. So frugal is she with all her bountifulness! So provident is she, and so well does she husband her resources! So good and careful a provider is she—never liable to be reckoned worse than an infidel because she does not provide for her own!

So well established is this principle in our experience that scientific men are accustomed to assume it and build on it without ceremony in their investigations, especially in physical science. As soon as they discover a constitutional physical want of any natural species, they at once receive a powerful suggestion of the existence, or at least of the attainability, of an adequate supply. Nay, they are profoundly convinced of such existence or attainability, and confidently assume it. If Cuvier finds a bone, he at once reconstructs the whole animal to which it belongs, and tells us how it looked, and what its habits were when living. How? On the observed fact that whatever is needed to complement a given mechanism in Nature and enable it to be turned to full account, at least in a natural class, exists or has existed. He does not speculate as to how this great fact came to be; only it is matter of plentiful experience that it is. — Mantell, in digging deeply into the earth, discovers a strange skeleton, and the round bony sockets where once were eyes. "This class of animals," he says, "though now lying five hundred feet below the surface, once had life above ground." "And how, my dear sir, do you know this?" "Why, do you not see that the animal had eyes, and needed to live in the light?" The philosopher would not be confident that some individual of the class did not spend all its days in darkness underground; but he is sure that it was not so with the class at large. They needed life in the light, and he sets it down as certain that they had it; and not a naturalist in Christendom will dream of disputing him.—As little will Owen be disputed, when, on finding on the surface of a field the bony frame of an animal which plainly never had any eyes to speak of, but which had feet and form and head nicely suited to living and making its way underground, he just reverses the former assumption. "This class of animals," he says, "needed life below the surface;" and on the instant he takes it for granted that the life which they needed they had.—Miller goes far inland, and there, on the top of a mountain, picks up the bleached debris of a bird plainly once web-footed. This class of birds needed, in part at least, a life in the water; and his mind at once rests unwaveringly in the conclusion that at the time when they lived a water-life was accessible to them.—Or Sedgewick, or Dana, or Agassiz finds a fossil with both herbivorous and carnivorous organs: only to be quite sure that the species to which it belonged, needing both flesh and herb for its best development, lived in times when both flesh and herb food could be obtained. Or, these eminent philosophers find a whole formation filled with fossil plants and animals having special adaptations to an amphibious life, needing a world of ponds and marshes, not indeed to exist, but to exist after their most flourishing manner: only to feel sure that those amphibians by structure actually lived in a transition period when the world of waters was just giving way to a world of dry land. As soon as they see the need, they believe in the supply. And from east to west of the scientific world there is no one to lift up a single sign of remonstrance. "The whole earth is quiet, there is none that moves wing or opens mouth or peeps." It is universally felt that this is sound science — that no sounder is to be had in all the Baconian realm. And so on to a vast extent. Behold scientists perpetually assuming and allowing that somehow supply accompanies the generic needs of Nature as shadow accompanies substance — sometimes before it and sometimes behind it, sometimes near and sometimes considerably removed, sometimes easily seen and sometimes seen with difficulty or not at all; but always existing, as surely as there is always more or less light on the earth even in the darkest night, and always linked to its counterpart substance by indissoluble though invisible bonds!

Now, this conduct in men of science does not proceed from a traditional notion of a wise and good God who will do nothing by halves or tantalize his creatures, but from a sense of what is the general course of Nature. It is observed that somehow Nature has a way of finding her generic wants supplied — that is the whole of it. It is observed that somehow her parts are extremely apt to orb themselves out into wholes — that is the whole of it. It is observed, and profusely observed, that somehow it is with her as with other kind and wealthy mothers; she does not send forth her children into the world without suitable outfit — that is the whole of it. This is the secret of that high scientific confidence. It is all pure observation — not at all traditional theology. Scientific investigations seldom proceed on theological grounds, even among religious men. Indeed, it has long been established law that special jealousy be used to prevent anything of the kind. Besides, the conduct referred to is as general among those who never think of a God in any practical connection with their employments — among atheists and antitheists who are carefully and zealously on the watch against any tacit assumption of the general Theistic tradition — as among others. Even as the great mass of farmers and merchants and other men of affairs never go behind natural agents and laws in the dealings and conceptions of their business, so with the investigators of Nature as a class — they deal exclusively with phenomena and natural causes. They see on all hands the immense tendency to equilibrium. They see how the streams and straws converge on every vacant spot. They see that wherever there is a broad organic need thither set gulf-streams and trade-winds freighted heavily with relief-ships. They see that where shines the Castor of a demand you may with proper search find shining over against it the twin Pollux of a supply; and that where night appears there appear also the festival-keeping stars. They see — as they have always seen from the time when they began to observe Nature at all — that she has a happy faculty, say a genius, at getting her loud calls affirmatively answered, her great hungers and thirsts provided for, her hungry vacuums charged with fitting contents. They see — in short, it is pure sight from beginning to end. Witness almost all the men called philosophers, from Comte upward. And from Comte upward, the idea of a Framer of Nature who is far too wise to frame useless things, and far too steady to His purpose not to carry thoroughly through whatever framing He has begun — this idea has nothing to do with prompting the general philosophic conviction that the great polarities of Nature are everywhere as faithful guides to explorers as when boxed up in the mariner's compass; and that when we find her putting up at some corner of her thoroughfares a guide-board, with its striking index-hand and great capitals saying "London" to all passers by, we may assure ourselves that London exists — more especially when she proceeds to plant her own person by that prophetic cross, and to glare with the eye, and to point with the finger, and to nod like Olympian Jove in the same direction, and to exclaim "London" in every civilized speech and with a voice that surges against the stars.

For see further. Physicists, on purely natural grounds, not only give unanimous consent to the principle that in their field Nature has an inveterate habit of getting her generic needs supplied, but they will, on the same grounds, unanimously consent to a certain extension of the principle — namely, the larger the need, the greater the momentum and evidence with which Nature finds the supply. There are four cases. Sometimes a need is intrinsically larger than another need. Sometimes several needs point at and demand the same object. Sometimes the class to which a need belongs is broader and more important than another class. And the strongest case of all is when you have a combination of these three cases into a fourth — when you have great needs, many of them, and all of these belonging to a class of immense breadth and importance. Then the aggregate need is very great; and the supply of that need is assured after a most manifold and imperial manner.

Some needs are intrinsically greater than others. And we observe that it is after the spirit and manner of Nature to give the most heed to the loudest call. See with what peculiar care she guards such vital things as brain and heart behind their bony ramparts! Whatever it is that prompts her to provide for a need rather than for a no-need, would prompt her to provide for a great need with more care than for a small one. — Sometimes several great needs point at and clamor for the same supply. As we have seen, each of these is evidence of the existence of that supply. And together they are so many independent evidences of that existence, and form an aggregate need of the largest dimensions. The manifold call is extremely loud and pressing. It must receive a correspondingly great attention from Nature. For such is the way of her who provides for a need rather than for a no-need, and for a great need rather than for a small one. — Some classes, each having several great needs, are more extensive and important than others. Make sure that the same notoriously self-consistent Nature that is everywhere more careful of a heart than of a hair, of wholes than of parts, of classes than of individuals, is more careful of great classes than of small ones. And if the class that clamors for a certain supply with many loud mouths is enormously broader than another class, then the sure testimony of Nature to the existence of the supply is enormously louder and more impressive. It is the voice of many waters. It is the concurring affidavits of many independent witnesses, each of which swears with a steady and determined voice. Hence, if we find all these cases combined in one — if we find very great needs clamoring for a supply, and many such needs all shouting for the same supply, and these all belonging to a class of beings enormously large — then we have a threefold assurance of the intensest and broadest character that the object for which so many brawny hands are stretched out, and so many brawny voices call, is extant or attainable.

For example. The class of men need air. This need is of a more crying and imperative kind than that for air of standard density or purity; and so Nature is more careful to secure the former than the latter. Also, it is not only the lungs of men that imperatively needs air, but also the ear and the nostril and the eye and the skin with its adapted pores — to say nothing of other organs. Neither of these can be turned to proper account without air. So it is a chorus of loud calls that is made for the same thing: and the whole is greatly more loud and impressive than any single call would be, and so proportionally surer of being heard. Also, each call is an independent argument for the existence of the supply. Further, it is not the class of men only that has these various needs; but the greatly larger class of living beings — comprising hundreds of thousands of animal and vegetable species, each on the average more numerous than men. All these would be sacrificed without air. This exceeding breadth of calling Nature gives exceeding volume to her voice. It rolls on the ear like tropical thunders. What place so remote as not to hear? What place so deaf? And every distinct organic species that helps to make that great invocation is a distinct proof that the invocation is not in vain. Altogether, the proof is immensely cumulative. It convinces like noonday mathematics. It displays more coöperative banners than Homer saw marching invincibly on Troy; or than Tasso saw waving and glinting above Godfrey and his embattled Europe, on their way to Jerusalem Delivered. So that when we find in the fossil realms of geology an immense amount and variety of plants and animals, all of which imperatively needed air, and all of which imperatively needed it for many purposes, philosophers feel that they move a million strong on the conclusion that when those races existed air existed for them. They would laugh to scorn the man who should pretend to doubt whether an atmosphere flooded the world in the Silurian or any subsequent age. They feel that nothing can be surer. And this, although it is simply an inference from the vast need of air — from the vast and varied waste that would be implied in case no air existed.

So in other similar cases. Wherever in physics we find many great needs calling for help in behalf of a prodigious sweep of being, we have an anvil chorus that does not fail of being heard. It commands like an emperor. It invokes like a mighty magician. It dredges the whole abyss of the unknown for an answer, as with Briarean hands; as a river is dredged for some lost favorite by a whole out-turning population; as all the declinations and ascensions of heaven have been dredged by our later Astronomy for hidden planets and nebulæ. O Polypheme, thou hast indeed a great and most successful voice! How the shores and the distant hills and the far, far away welkin ring as thy massive lips part skyward, and thy huge swollen chest empties itself in gales and torrents and cataracts of sound! And see how quickly the invoked help swarms to his aid! Swart forms as huge as his own, great voices as burly and earth-quaking as the voice that calls, make prompt answer, and say, Here we are! The Cyclopean voice has found out the Cyclops.

"Clamorem immensum tollit, quo pontus et omnes
Intremuêre undæ, penitusque exterrita tellus
Italiæ, curvisque immugiit Ætna cavernis.
At genus e sylvis Cyclopum et montibus altis
Excitum ruit ad portus et littora complent."

Such is an uncontested law of Nature. Each real need of a natural class of beings has over against it in Nature the suitable supply, or the means of it; and the greater the need, the greater the momentum and evidence with which the supply is furnished. We have seen that modern science is built on this law to a very large extent. We have seen that even atheistic explorers of Nature would be glad to build any amount of science on the same foundation. And shall any complain as we now proceed to build upon it what is of more concern to the world than all the profane science that ever was taught and cried up to the third heaven of fame — even though it unravels the tangled mystery of the stars, and provides all the dynamics of useful industry, and probes the solid earth to where sleep in their stony mausolea the secrets of her genesis and hoariest history — I mean that sacred science, the Doctrine of God!

For God is needed. I do not claim that He is needed to keep the frame-work of Nature — its mountains, plains, and seas; its plants, brutes, and men — from promptly falling into utter nothingness: nor do I care to say that every particle of order and comfort would at once drop out of the system were an active Divine hand withdrawn from beneath it.

Still God is a necessity to the universe. Human nature in its soundest state hungers for Him. The race at large needs to believe in Him. The total system needs to have Him — as a Providence and Government. Woe to the morals of mankind when faith is wholly dead! Woe to the happiness and usefulness and all true greenness of mankind when its morals have gone down into the Potter's Field where faith lies scantily buried! Woe to all earthly Nature, animal and vegetable as well, when the human life and character have both become offensive corpses! Nay, woe to the universal Cosmos when it has no God to guide and govern it! An infinite evil has come upon it. It has sustained unspeakable loss. For suppose God to be *dead.* He has lived and wrought and governed as only a being of perfect faculty can, through amazing æons. But now He dead — dead. What means such an event to the universe? Manifestly it means something very dreadful. The difference to the system between the presence and the absence of such a dynamic as God is plain infinity. And so I declare that an unspeakable loss has been sustained. A good that defies figures, and even the fancy, has been subtracted from the creation. It is not too much to say that the creation became bankrupt at that loss — became, at that stroke of doom, both altar and sacrifice; a holocaust sacrifice, whose lurid flames make all space ghastly, and go on to fill it with the charred and ruinous heaps of its former fair self. Oh, how these dismal ruins hunger and thirst for the old God! Oh, how much these black and cindered earths yearn and beseech after Him through their innumerable fissures and gaping wounds, as the parched and chapped ground in fiercest drought yearns and cries toward the heavens for rain! At last true midnight has come. And from out its anguished bosom, indestructible though ruined Nature sends forth groans on groans accented with profound despair — broken with piteous protests and pleadings for a God that cannot be spared. There is not a scorched and scarred fragment that does not, forgetting all other wants, join in the mournful chime, Oh for a God, Oh for a God! The state of things is such as to invoke the dead God from His nonentity with almost the force of a Creator. And should He, in answer to these appeals "creating a soul under the ribs of death," again suddenly make His appearance — appear no more to die — Oh, what thrills would circulate, what hallelujahs would go up, what jubilees of shouts and songs would peal and re-peal, what an ecstatic wave of sacred laughter would run and flash in the new sunlight across the whole breadth of being! Nature would ring all her bells. She would blow all her silver trumpets. Even demons, methinks, would rejoice as they emerged from that chaotic night into morning. Even they would be glad at that greatest of Eurekas — the refinding of One who knows how to govern. Never before has such exulting Sabbath been kept — not even when the morning stars sang together, and all the sons of God shouted for joy. And it is but just. So utterly measureless is the need of God. No ordinary words of literal statement can begin to carry a just sense of its greatness.

I saw a little child — wandering, wandering. A strange place, new objects, fresh curious eyes peering at everything—little one, where are you going? Suddenly she misses her father. Where is he? She looks round and round. Where is the familiar hand that a short time ago held hers; where the familiar face that but just now was looking down upon her so tenderly and protectingly? A great fear begins to steal upon her. She lifts up her voice. "Father, *Father*." No answer comes. "Father, *Father*, FATHER." Still no answer. Her alarm increases fast. She begins to run about and to ask of one and another, with flushed face and anxious, questioning eyes, "Have you seen him—Have you seen my father?" No — no one has seen him; and the poor child's heart fails her more and more. Her knees begin to tremble, she is all agitation, her questionings and calls become every moment more hurried and tremulous; at last a spasm of mingled wails and calls pours out of her white lips, and she breaks down into convulsive sobs which no soothings can allay. Poor child — she has lost her father! Will she never see him again? God forbid — if there be a God.

This partly expresses what every good man would feel like doing on discovering himself to be without a God — what all men, after a short experience of doing without Him, would feel like doing — indeed, what comprehensive Nature, could it become personal, would certainly do. She would, with a great fear at her heart, begin to call for the lost Heavenly Father. She would go searching for Him and asking for Him up and down all the latitudes and longitudes. And should she not succeed in finding Him; should no answer come to her loud and urgent invoking; should she nowhere meet with any who could tell of having seen Him or His like; should at last the deepening gloom and silence whisper He is *dead* — oh, what a heart-breaking wail would go up that moment to pierce the very heavens! Drenched in sobs and tears, that poor orphaned child, though bearing the great name of Nature, would wish herself dead also. But hark — what sound is that? It comes nearer and nearer. She has caught it. She lifts her streaming face and breathlessly attends. On comes the strange sound — deepening and widening and at last taking on a perceptible accent of joy. She springs to her feet. With parted lips and face pictorial with hope she leans toward the advancing murmur. "He is found," "He is found" — at last shapes itself distinctly to her greedy ear. Lo, the day breaks like noon over her beaming face. She springs forward, she runs, she leaps, she meets Him afar, she clings about His knees, she casts herself into His open arms and nestles in His bosom. Her sobs die away. She looks up into His face and smiles and clings, and clings and smiles. At last she is at rest. God bless her — at last the poor child whom some call Nature, is at rest. And the tears of by-standers — the spaces and durations — fall in joyful rain. It is all right — just as it should be. It would have been such a dreadful orphanage! Such rags, such hungers, such rooflessness and homelessness, such neglects and exposures and blows, such ignorance and guilt and misery — never were seen the like. Poor Nature would have DIED.

What is the scientific import of all this? This need of a God, so pronounced and mighty — what means it in view of such science as forms the glory and boast of this nineteenth century? Why, it means the same as every other case of natural generic need — it means a supply of the need; it means an actual God. Nay, it means Him with an emphasis more prompt and sonorous than ever came from any other need whatever. For it is the supreme need. It is the Labarum among standards, and the Pontifex Maximus among priests. Indeed, it is much more than this. It is not only a supreme need but an incomparable one; not only an incomparable one but one which could not be greater. How could a need cry more out of the very depths and essential nature of finite things? How could it relate to more vital things than virtue, and those forms of happiness and usefulness that naturally follow in the wake of virtue? How could it bear on a greater variety of interests than Infinite Faculty can reach and aid, or bear on them more heavily? How could it spread itself over a wider area of being than the enormous All? So that it is not enough to say that Nature aches after God more than ever eye ached for light, or lungs for air, or wings for an atmosphere, or fins for waters, or the babe for its mother. It is even not enough to say that this Divine need towers out of sight above all others; that it is indefinitely beyond rivalry; that, *toto orbe*, there is nothing else for which Nature clamors so deeply, from so many mouths, with such variety and breadth of voices, and in behalf of so many classes of being. It is not enough to say even this. We must go on to say with plenary voice that it is not within the range of possibilities for a need to be greater at any one of these essential points. It is a maximum of maxima. It is the eternity among the durations, the astronomical abyss among the spaces, the God among living beings. The day that should see the universe vacated of a God would begin to see it beggared to the last farthing. And when we have said all this, and truly said it; when all such secular needs as we have been instancing appear dwarfed into nothingness by its side; I demand in the name of Science — Shall all these little needs, without exception, be allowed to argue a supply for themselves, and the same privilege be denied to this greatest? Shall all these be allowed to argue in proportion to their size, and it be denied to this need to argue at all? Science protests against such strange proceedings. She will not permit such huge inconsistency — especially as done in her name. After we have taken indefinite consecutive cases of organic and generic need, and found them always balanced by supply with a brilliancy and emphasis proportioned to their size, shall we come to the next case — which happens to be that of the Divine need — without expecting a continuation of the law? Who will be so unscientific, unreasonable, and absurd? Who will not feel able to rest all his weight on that long chain of inductions? What scholar in the Baconian philosophy and the philosophy of common sense will hesitate to declare that this need, which transcends all others so immensely in magnitude, transcends them correspondingly in the force with which it argues for that grand supply whose name is God?

So the *nisus* predicts a God. The Being who is needed to complement Nature into a whole really exists. The great hunger for Him which belongs to normal human nature is provided for; the great vacuum that beseeches Him like a maelstrom can be filled; the great finger-post that points at Him so steadily from all corners is not perjured; the great drafts and streams that set in from all quarters toward Him really have Him for their Equator; the great wheel whose radii are seen converging on Him till they are lost in clouds really has Him for its sublime Axis.

See how these flowers, and indeed all this abundant vegetation, have their generic slant sunward! Do I need to have seen that attracting sun during all these months and years in order to know that it has existed and shone? Nature is a heliotrope — an enormous sunflower, turning its whole fruitful bosom toward God; and when I see that generic bent, I do not need to see God in order to know that He is. I see the obeisance which the creatures are paying to their Creator. — See how these vines lean and twine and cling and put forth their profuse rootlets and tendrils! Does one need to see the firm trees and sturdy walls and century-defying church-towers to which these epiphytes fasten themselves, in order to know that such supports may be had? Nature is an ivy — a leaner and clinger by its very structure, with tendrils and rootlets innumerable issuing from all parts, and reaching for support and nourishment to something indefinitely firmer and richer than itself; and I do not need to see cathedral God in order to know that this glorious support for Nature may be found. The very structure and infinite tendrils of that wonderful creeper proclaim Him. — See how mysterious instinct draws the babe toward its mother the bee toward its cell-building and honey-making, the silk-worm toward its spinning, the coralline toward its submarine architecture, and each species of living Nature toward its peculiar functions and line of life! Surely, if one could know these instincts apart from the things to which they point, he would not need to actually see that mother with his two eyes in order to know that she exists — or that curious honey-comb with its plenum of mathematics and nectar; or that cocoon wealthy with the silks of Lyons and Cathay; or that coral archipelago within whose harbors navies safely ride, and on whose fertile bosom tropical harvests bloom and empurple — these things are all implied and sworn to in the very instincts themselves. Such, from babe to coralline, is Nature; and the true Mother, the infinite Sweetness, the gorgeous Robe, the tropical Paradise to which it instinctively reaches forth and calls, is God. Why must I see Him in order to know that He is? The very instinct that blindly draws and pushes everywhere toward utility and beauty and goodness and worship announces Him sufficiently.

See Uranus wavering and quavering on his Siberian path. Must I put a telescope to my eye, and descry perturbing Neptune, before I send in to the Institute my account of the new planet? It alone satisfies the perturbations. Still look, O German Galle, and all ye whose faith in mathematics and the law of gravitation is weak; look toward Delta Capricorni, and optically find what is already theoretically known. — See all the path-bits of the solar system curved as for a common center, and lo, some of the celestial pilgrims brightly smiling toward the same point! Who feels that he must actually see that center blazing as a sun before he can solidly believe in it? Why, all the arcs of the system, great and small, unite in affirming that primate and metropolitan.—See Constellation Hercules growing larger, year by year. Must you see, with fleshly eyes, a flaming ellipse trending along the abyss, and carefully take its bearing among the stars with compass and sights, before you will consent to believe in it? If so, alas for the Herschels and Struves! They are visionaries, and not the men of science they have had the credit of being.—See the proper motions of all Galacteal stars curved as if for central Pleiades! To know the reality of that center, must I actually see it blazing like twelve thousand suns, and actually see it brightly zoned about by its eighteen millions of completed ellipses, and actually hunt down, one by one, as many shadowy foci till they are lost to view in thy effulgent bosom, O illustrious and imperial Alcyone? Not at all. Forbid it, Dorpat and Pulkova—forbid it, the fames of Mædler and Argelander and all most signal astronomers! Never do I need turn eye on the neck of Taurus. Its famous cluster might be as strange to my sight as the lost Pleiad. And yet I must believe. It is enough for me that I know the law of gravitation, and have noted the general drift of our heavens. This settles the matter. Every bit of star-path out in yonder vault contributes a voice to that euphemism which tells me the brilliant story of the Central Sun. I am assured of that august nebular heart, of that astonishing center of force and revolution, as plainly, if not as impressively, as I could have been by near sight. No, I do not need to see it. No more do I need to see God in order to know of His existence. He is perturbing Neptune. He is the Herculean Constellation toward which all things sail. He is the metropolitan Alcyone around which all things revolve. So I have no occasion to invoke sight. The perturbations of Nature show Him. Her orbits concave to Him proclaim Him. The general drift of her firmaments announces Him like a choir of trumpets and artilleries. Hail, Great Center of revolving being — as real as if we saw Thee on Thy throne sending forth Thy beams and government to remotest space! The *nisus* has revealed Thee; and it was not in vain that we adjured

> "Per magnos, Nisu, Penates
> Assaracique Larem, et canæ penetralia Vestæ
> Obtestor; quæcumque mihi fortuna fidesque est
> In vestris pono gremiis: revocate parentem;
> Reddite conspectum; nihil illo triste recepto."

VIII.

THEISM

AS A

SCIENTIFIC HYPOTHESIS.

Ὑψώθη ὑπὲρ παντα, ὑπερωμίαν καὶ ἐπάνω.

Χρύσεια τάλαντα—Τρώων χῆρες πρὸς οὐρανὸν ἄερθεν.

Homer.

EIGHTH LECTURE.

THEISM AS A SCIENTIFIC HYPOTHESIS.

I ASK your attention in this Lecture to the superior merits of Theism as an hypothesis for the explanation of Nature. Notice that the hypothesis, while perfectly sufficient, and, to say the least, *a priori* as credible as any, is vastly the simplest, the surest, the safest, the sublimest, and the most in accord with the convictions and traditions of mankind, especially of the most enlightened and moral part of mankind. Some of these particulars may appear at first sight to address themselves solely to the taste and interest: I trust they will be found to appeal to the reason as well.

The Theistic Hypothesis is perfectly sufficient. It is plain that a Being of power and wisdom indefinitely beyond the human can completely account for all the wonders of Nature. Nothing could be plainer. A child can see it as well as the sage. The most exquisitely fashioned man, the noblest astral system, the aggregate of the amazingly varied organisms that crowd the earth and spangle the sky — a God is abundantly equal to the production of them all. There is absolutely nothing which such a Being cannot do with the greatest ease. He has skill enough to contrive the most exquisite things, power enough to accomplish the hardest things, and comprehension enough to triumph with these attributes over the largest fields of being which observation has examined or thought conjectured. As an explanation of Nature, the Theistic hypothesis could not be improved. The hardiest assailant would scarcely dare question its perfect sufficiency.

The Theistic Hypothesis is, to say the least, a priori, as credible as any.

The various hypotheses to account for organic Nature are as follows. First, natural organisms, as individuals or races, are eternal. Second, they were constructed by chance. Third, they were constructed by law — that is, by blind material elements acting in obedience to the eternal laws of their natures. Fourth, they were constructed by God.

The first two suppositions are too openly in conflict with observation and science to find any supporters in this age. No one now supposes that the individual plants and animals which he sees about him have always existed as such. That tree, that brute, that man — each of these individuals self-existent, imperishable, eternal! All the senses of all men are against it. They protest in a thousand ways that such organisms begin and flux and dissolve with every passing day. Equally plain is it that the races began — as plain as the igneous and metamorphic rocks, and the alphabet of geology. — As to a man, or even a blade of grass, becoming constructed by a strictly fortuitous concourse of atoms, such epicureanism is now out of date by many centuries. Chance — no person of culture at the present day believes in such a thing! Nor is it Argyle alone who believes in the reign of law. The schoolboy or the schoolless peasant does it as well as the cultured noble. All persons among us now understand that every atom has its essential properties and laws, which, together with those of other atoms and agents, spiritual and other, determine all its doings and experiences. The very idea of hap-hazard died and was buried at the incoming of modern science; and every new inquiry into Nature heaps new measures of dust on the grave. So we may put aside the two hypotheses first named. The comparison lies wholly between the last two — between that of construction by law and that of construction by God. Which of these has the best claim on our favor?

Let it be observed, in the first place, that the hypothesis of construction by God is, to say the least, fully as credible, on its face, as its rival. Of course a person is perfectly credible; for we know millions of such beings in actual existence. Of course a person producing organisms, and very elaborate organisms, is perfectly credible; for we know millions on millions of persons actually doing as much. An eternal and practically infinite personal constructer of organisms is a more difficult conception, and further removed from our experience; but not more so than the eternal and practically infinite material constructer of organisms which the law hypothesis assumes; for it assumes what is really a material God — an eternal assemblage of blind atoms with properties in the aggregate fully equivalent, so far as production of results is concerned, to that personal power and wisdom indefinitely greater than the human which the Theistic hypothesis ascribes to God. Indeed, the construction of organisms by an intelligent agent is wonderfully more conformed to experience, not to say reason, than the construction of such organisms by mere blind matter. We have no conceded instances of the latter construction, while we have innumerable conceded instances of the former. Men do plan and execute watches, telegraphs, sewing-machines, pin-machines — machines beyond count. This is matter of absolute knowledge. It is universally granted among those who believe in knowledge at all. But construction by mere blind force, is not granted — especially construction of intelligent and moral beings. Only a very few imagine such a thing proved at all, and they in only a few instances, and that rather as a possibility or a presumption than as a demonstration. And just think of it. A mist of blind elements blindly shaping itself, not only into an infinity of useful and admirable objects — and such only — like plants and animals, but also into intelligent and moral beings; into statesmen, philosophers, and saints; into Napoleons, Miltons, Newtons, Howards; in fine, into such books as the Principia or Paradise Lost — for the author cannot be less wonderful than his works. What says an unsophisticated mind to the idea of matter, under blind forces and laws, shaping itself into the Iliad, or the Mécanique Céleste, or the mosaic portraits of the popes that look down so marvelously in long order in the Roman St. Paul's? Why, the very idea gives a shock to the understandings of most men! It seems like an insult to their intuitions. It seems to defy their common sense and knowledge of Nature. Blind causation do such things! To say that the conception is hard, far-fetched, unnatural, is not enough. It looks vastly preposterous. It begs like Demosthenes to be considered a self-contradiction. The man who accepts it instead of Theism, has wonderfully the appearance of one swallowing a camel after straining at a gnat. Blind causation do such things — it seems a feat a hundred fold more wonderful than any ever attributed to a personal God! Really, the hypothesis of construction by law is, on its face, greatly less credible than that of construction by God.

The Theistic Hypothesis is vastly the simplest.

Each hypothesis, considered as an explanation of Nature, consists formally of two parts — first, certain assumptions; and, second, certain considerations to show that these assumptions, in connection with known principles, will explain Nature. In the case of the Theistic hypothesis, the first part consists of the supposition of an eternal Being with power and wisdom indefinitely greater than the human, while the second part is nil—no considerations whatever being required to show that such a Being can account for the whole hight and breadth of Nature.

Not so with the law hypothesis. Here the two parts are much less simple, being in fact two very generic and complicated schemes of suppositions and arguments; one called the cosmical hypothesis for explaining the origin of worlds, and the other called the physiological hypothesis for explaining the origin of the living organisms on this world. The leading suppositions of the general scheme are as follows:—

1. An eternal substance, namely, matter.

2. An infinite number of eternal substances, namely, countless material atoms having independent existence.

3. An eternal and infinitely complex scheme of exquisite relationships between these countless, eternal, independently subsisting substances.

4. These exquisitely correlated atoms tenuously diffused as a gas or mist.

5. This mist vastly larger than a solar system.

6. This mist on fire.

7. Currents in this mist, obliquely toward the general center of gravity and nucleus of condensation.

8. Several minor nuclei of special condensation distributed through the mass — each with its own system of oblique currents.

9. All these nuclei such in size, place, and number, as to harmonize with the conditions of stable equilibrium in a solar system. I call particular attention to this last most voluminous assumption.

These are only a part of the assumptions included in the law hypothesis — merely leading specimens. You observe that the infinite and eternal enter quite as largely into this scheme of explanation as into the other — indeed, more largely — while there is no comparison between the two schemes as to number of assumptions.

But, allowing these numerous assumptions, it does not intuitively appear from them, as it does from the assumptions of the Theistic hypothesis, that they will explain Nature. Arguments are necessary. No small amount of them is necessary. The arguments to show that the foregoing postulates, with the help of known laws of matter and principles of science, are adequate to explain natural organisms, may be arranged in three classes: —

First, certain arguments to show that all the worlds composing our solar system, and the leading features of each world, may be naturally derived from the foregoing data. These arguments are long and intricate; and when duly spread out, make a volume.

Second, certain arguments to show the possibility of spontaneous generation of the lower forms of organic life. These arguments are long and intricate; and when duly spread out, make a volume.

Third, certain arguments to show the possibility of transmutation of species by gradual natural development of these lower organisms into higher forms, and at last into intelligent and moral beings. These arguments are long and intricate; and when duly spread out, make a volume.

Now, granting that the two schemes of explanation are, in the last result, equally good at accounting for Nature, you observe that one is a vastly more complex plan of explanation than the other. With elements fully as difficult, one consists of many parts — the other of few. With elements fully as difficult, one requires volumes to unfold itself fully — the other requires only a few words. Need I ask which is the more philosophical? It is an immemorial and indisputable canon of philosophy to accept the simplest explanation of facts.

We have taken the law hypothesis in its usual form. If any one thinks it may be made more simple by supposing more and arguing less, let him try. Let him reduce the second part of the hypothesis to zero by introducing the following comprehensive supposition into the first part. Suppose those eternally correlated atoms to have an efficiency practically infinite — to have forces and laws which as a whole are fully equivalent, so far as results are concerned, to that power and wisdom indefinitely greater than human which the Theistic hypothesis ascribes to God — to have forces and laws which are of themselves able to bring the atoms together into all the exquisite organisms that we see, up to intelligent and moral beings.

Of course, to grant this supposition is to grant everything. No need of any argument to show that such an hypothesis will explain Nature. But is such an hypothesis plainly allowable? Does it assume only what is plainly possible? All the assumptions of the Theistic hypothesis are assumptions of what in the nature of things are evidently possible — an eternal person, this person indefinitely superior to man in wisdom and power. But this last assumption of the law hypothesis is a very different matter. To take for granted that a mist of atoms, by virtue of any blind properties whatever, can arrange itself into that infinite variety of exquisite organisms — and nothing but exquisite organisms — that we see, is taking for granted a great deal; is taking for granted what one may well be pardoned for doubting. The possibility of such a thing needs mightily to be shown. It needs to be shown that astonishing solar systems can result from mere natural forces and laws; that spontaneous genesis of organic life in some low form can occur; that there may be a natural development of this low form, through transmutation of species, into the most wonderful men. The possibility of all this needs not only to be shown, but to be shown to a demonstration; since all the assumptions of the rival hypothesis are possible to an absolute certainty. It is self-evident that there is some eternal substance, and that an eternal person is, in the nature of things, just as possible as eternal matter — self-evident that there is nothing in the nature of things to limit an eternal intelligence to a given breadth of knowledge and power — self-evident that there is nothing to prevent that intelligence from being as much greater than men in these respects as man is greater than a worm. Thus in the Theistic hypothesis. So everything in the rival hypothesis must be put on a basis of absolute certainty. That profuse argument, drawn out through volumes, which undertakes to show the possibility of a cloud of blind atoms doing the work of an infinite God must be made as strong as Euclid. Every link in that long chain of evidence must be forged by some Vulcan in the smithy of geometry. On this plan of exhibition, the law hypothesis will be quite as complex as on the other plan. On both plans it is a most cumbrous machine for its purpose — wheels within wheels in most unnecessary and perplexing maze. It is the first rough effort of the inventor compared with the instrument when, at last, simplified into a tithe of its original size and expense by the labors of many years and many rival ingenuities. It is the long, rambling, tedious process of some unfledged geometer compared with the swift and laconic algebra of La Grange. It is the Ptolemaic system of Astronomy compared with the Copernican — the vortices of Descartes compared with the Newtonian principle and law of gravity. What manufacturer now uses the first spinning-jenny of Arkwright? What mathematician now works at his daily investigations with the ancient synthesis rather than with the modern analysis? What astronomer now explains the heavens according to Ptolemy? Cycles and epicycles and deferents and eccentrics piled on each other — who does not bless himself that he is well out of this tangled wilderness into the grand simplicity of the Copernican theory? With a true philosopher, nothing but the clumsy manifoldness of the old system, as compared with the new, is needed to secure its emphatic rejection. Could it explain all astronomical facts equally well with its simpler rival, it would still fail of countenance for a single moment, as being essentially unscientific. So should fail of countenance that complex and cumbrous law hypothesis which is the Ptolemaic system of natural theology. However successful it may prove in accounting for Nature — though it should leave nothing to be desired in respect to clearness and certainty of result — it ought to be summarily rejected as being a tedious Chancery and Circumlocution Office. What traveller rides with a fiftieth or even a fifth wheel to his carriage? What American, seeking merely New York, goes by way of Pekin?

The Theistic Hypothesis is vastly the surest.

It is perfectly certain — certain to the apprehension of all mankind — that the hypothesis of a God will account for all natural wonders.

Can as much be said in favor of the hypothesis of construction by law? Is its adequacy intuitively certain? Or has that adequacy been rigorously demonstrated, level to the apprehension of all the world? No one claims it. No one dares to claim it. Great effort has been made. Great ingenuity has done its best. Years of argument have piled themselves on years, and still the argument rages. With what result? The great majority of thinking men are as unconvinced as ever. They do not even find a modest probability in the scheme so laboriously commended to them. And even its best friends hardly presume to call their own arguments a proof, much less a demonstration, much less still a demonstration that can be universally seen to be such. A certain amount of philosophic credibility, or, at the most, probability, is all that such men seem to themselves to have accomplished by their long and intricate dealings in behalf of spontaneous generation and transmutation of species by natural development; while to most persons the whole scheme is a hopeless fog-bank — very picturesquely constructed perhaps, and displaying not a few showy battlements and pinnacles and prismatics — but still mere unsubstantial and uncertain air-castles, liable to change shape and even disappear at any moment. And yet, to put their scheme on as good footing as the Theistic, its ability to explain Nature must be made a matter of absolute and immeasurable certainty to the gaze of all plainest understandings. For, from sunrise to sunset, and round to sunrise again, there is not a person capable of understanding the proposition who does not know, to absolute perfection, that an Infinite Person could produce with perfect ease the noblest and all things that make up the beauty and majesty of Nature. It is as much an axiom to the child and the savage as it is to the sage. So a heavy demand is made on the friends of the law scheme. It is not enough, should we find ourselves unable to prove positively that this scheme is insufficient to explain Nature : its friends must show to utter certainty that it is sufficient, and show it to the complete satisfaction of all respectable inquirers. A hugely contested probability, timidly accepted as such by a few respectable reasoners, will not answer. Euclid himself must not be more conclusive, nor his axioms plainer. To secure this, all the parts, scores in number, of the very complex scheme, must be put on the footing of geometrical axioms. You must do it for all parts of the cosmical argument. You must do it for all parts of the physiological argument — for the spontaneous generation, for the transmutation of species, for the development of the oyster into the Newton. Not a single point in the voluminous scheme must be left to rest on mere probability. Should absolute demonstration halt at a single one of these points, or at any one of them fail to flash conviction like a sun on the most limited of sound understandings that chances to glance thither, the whole hypothesis would break down as a demonstration. Of course such a Cesarean, all-conquering proof, is not only unaccomplished, but unaccomplishable. Not an instance of it can be found in the whole kingdom of logic.

A man who, reduced to choose between two secular hypotheses in other respects equal, should choose the one whose adequacy to account for the facts is, almost unboundedly, the most questionable, would not be considered the wisest of men. Suppose you meet an English friend in yonder street. "How came you here?" you exclaim. He informs you that he came either by steamer or by artificial wings. Have you any difficulty in choosing between the two explanations? You can decide the case swift as the flashing light, and with the momentum of a planet. And why not? You certainly know, as does everybody, that a steamer is adequate to bring the man across the Atlantic; but you do not certainly know that artificial wings can do such a feat. Very far from it. What you know is that the possibility of such a mode of transit for men is extremely doubtful, to say the least. Some ingenious things can be said in its favor — witness Rasselas — but to most persons the very idea is very ridiculous, and to none is it more than plausible. So you have not a shadow of hesitation. Instantaneously, your mind flashes its decision. Between the hypothesis whose adequacy is perfectly certain, and the hypothesis whose adequacy is, to say the least, extremely uncertain, you have no occasion to linger. You take the immeasurably surer hypothesis immediately and as a matter of course. Your friend did not, Dædalus like, transfer himself across the seas by means of a pair of wings deftly fastened to his shoulders.

The Theistic Hypothesis is greatly the most salutary and safe — salutary for the present life, and safe as to another.

It is easy to see that the recognition of a God, carrying with it, as experience shows it generally if not always will, a recognition of His righteous government — certainly of the possibility of it — has greater tendency to restrain from misconduct and to stimulate to virtue than has atheism. This from the nature of the case. And experience accords. It lies on the very surface of life and history that Theism is better than atheism for the character, the happiness, and the general outward prosperity of communities and families and individuals. Such has been the teaching of my own observation and reading on this point, that I am free to say that I had rather have my child worship in faith some respectable Brahma or fetich than to have him altogether without a God. So felt the ancients, though with but a small part of our experience. Plato would have atheists exported far from his republic as being a public danger. He would have their children taken from them, and brought up as orphans at the public charge. And the words of Cicero to the same effect have become famous. "That such views are useful and necessary, who will deny, when he reflects how many things must be confirmed by an oath, how much safety there is in those religious rites that pertain to the solemnization of contracts, how many the fear of Divine punishment keeps back from crime; in short, how sacred and holy a thing society becomes when the immortal gods are constantly presented both as judges and witnesses." So spake classic antiquity. And modern times, with their larger scope, venture to speak still more strongly. To them Theism is like a certain geode but recently found. To them Theism is like a certain flower just now becoming naturalized in our conservatories. The stone was broken, and lo, it was lined with beautiful crystals, and in the heart of that rich casket a still richer crystal in the form of a cross! Some delicate petals of the Flos Sancti Spiritus are drawn aside, and lo, nestled in that fragrant bosom, looks forth what seems a milk-white dove! Such are the contents and implications of Theism — things most fair and wonderful to see. Behold altars and homes and commonwealths — behold orders, proprieties, safeties, philanthropies, steadfast consciences, regulated freedoms, and durable civilizations — behold usefulness and happiness and hope and virtue in their most snowy and effulgent forms — behold, as I think, the Cross and the Holy Ghost! All these are seminally contained in the Doctrine of God. It travails in birth with these for all the worlds.

Whichever hypothesis is honestly accepted will be measurably acted on. If that of a God is accepted, experience shows that with it, in general if not always, will be accepted His character as a righteous moral Governor. Supposing men to act on the supposition of such a God, it is certain that no grave harm will come of the action in any event, while it may open on the soul the gates of eternal life. But if men act on the supposition of No-God, they may be ruined remedilessly in case there is such a Being. Nearly all theists claim it will be so: a very plausible revelation affirms and reaffirms the claim in the most positive manner. And certainly, very severe results are by no means improbable. For, if there is a God, it is exceedingly important that men should know it; and if He is righteous — as certainly is not improbable — He greatly desires them to know it, and has given them suitable means for knowing, and so will be severely displeased with their atheism.

It would obviously be irrational to choose the least useful and safe of two hypotheses in other respects equal. No man in his senses would advise such a step in secular matters. It would be alike an insult to interest and to truth. I say, it would be a libel and outrage on truth — that Divine principle which is only inferior in beauty and majesty to virtue itself, and which is universally allowed to deserve the love and homage of mankind. Usefulness and safety are near of kin to truth. They are its natural associates. Where they are found truth is likely to be found. They are the surface indications of the gold mine — the Geology that divines of it so strongly that men hopefully gather great capital about the spot where trembles her rod, and set to work. If observation shows anything, it is that the most salutary and safe course is usually the one accommodated to fact: and indeed such a course cannot in general be that which is accommodated to a falsehood. From the nature of the case, courses accommodated to a falsehood, and so in positive conflict with the real nature and relations of things, must in general be attended with more difficulty, expense, and damage than those in harmony with such nature and relations.

The Theistic Hypothesis is greatly the fairest and sublimest.

Other things being equal, the fairest and sublimest hypothesis has the best claim on us — on our faith as well as on our affections. It has most the aspect of a truth.

Soul — whether regarded as an immaterial substance, or simply as the sum of certain qualities occasionally found in connection with certain organic forms — soul, with its will, feeling, intelligence, and capacities for happiness and virtue, is universally felt by thinking men to be the highest as well as the most mysterious sort of known being. Not the grandest masses of matter; such as mountains, oceans, stars — not the most subtle and forceful material elements ; such, for example, as produce the phenomena of light, electricity, and gravitation — not any conceivable combination of such elements, can compare in wonderfulness and nobleness with the soul of a Newton. Much less can any conceivable combination of such causes compare in these respects with an Eternal and essentially Infinite Soul that devises and produces all natural organisms, and is capable of governing them and all things with infinite wisdom and goodness. If, in addition, we suppose this great Being crowned with the glories of an infinite and everlasting actual felicity and virtue — as we are entitled to do for aught that appears to the contrary — a goodness efflorescing into every imaginable beauty of hue and form; a goodness bathing the whole Divine Nature in the rosy lights of an unutterable tenderness and mercy and love, whose warm floods overflow to the remotest terms of the creation, and insure to it the utmost possible measure of blessed results — what shall we say of such an Object ? It makes the heart leap to look toward it. Never such a scene blushed under eye of traveler or pencil of master — never such sumptuous palace or cathedral reared its wilderness of comeliness and majesty on the sight or dreams of men — never such mountain-range gathered clouds and rainbows about its brow and blossomed o'er all its mighty sides with the beauties of every clime — never such central sun blazed and triumphed and governed amid its coronet of rejoicing worlds! O wonderful Vision, O Colossus of perfection, O worthy and worshipful Emperor of Nature, O fairest and sublimest Idea in the whole empire of thought! One may well be excused for preferring, other things being equal, such an hypothesis as this. It has the most claim upon him. What should we think of a man who, being reduced to choose between two hypotheses equal in every other respect, should choose the meanest and hardest-favored of the two? It were an insult to truth. It would do violence to the subtle instincts and proprieties of Nature. It would affront the "beautiful and fitting" of science.

The Theistic Hypothesis is greatly the most in accordance with the convictions and traditions of mankind, especially of the most enlightened and moral part of mankind.

You could almost count up on your fingers the men who, leaving the attitude of mere doubters, have come to positively affirm and positively believe that Nature was actually produced in conformity with the law hypothesis. On the other hand, those who so positively and firmly believe in the Divine origin of Nature that they could freely die for their faith are almost innumerable. I would like to see the man who could die for the law hypothesis! — Further, the Divine origin of Nature is the strong popular faith of whole nations and generations, constituting the most intelligent and best-behaved part of the race. Much of this faith, indeed, is not that of martyrs; but most of it is a faith that shudders at the very name of atheist, and at the very idea of a godless universe. And the Jews, the Christians, the Mohammedans, the Hindoos with their affiliated races — to say nothing of smaller peoples — the believing nations covered by these names include in their mighty circumference nearly all the science and civilization and semi-civilization and respectable morals the world can boast. — Further, the whole body of mankind, past and present, with a few trifling exceptions, firmly believe in at least one Great Intelligence of a grade indefinitely superior to the human and worthy of worship. Every nation has some divinity. There is no country without temples, altars, priests. In all climates, under all governments, through all stages of society from the most barbaric to the most cultivated, man humbles himself before great invisible personal powers. The traveler into unexplored countries about as much expects to find them supplied with deities as he expects to find them supplied with men. The traveler into distant ages, whatever direction he takes, about as much expects to find men worshipping as he does to find them eating and drinking. Whether Livingstone or Humboldt — he encounters the supernatural at every step. Whether Niebuhr or Muratori — at every step he meets the immemorial traditions of the supernatural descending upon him like Amazons from every point of the compass. The cultus is everywhere. And whether it points at the fetich, or the idol, or the star, or the Grand Lama, or Brahma, or Boodh, or Odin, or Osiris, or Jupiter, or Allah, or Jehovah — it expresses the faith of all nations and ages in at least one Great Superhuman Intelligence who holds sanctuary within such holy names, before whose power and wisdom the greatest of men should uncover, and from whose undefined and dreamy greatness one should not be surprised to see issuing any conceivable wonders. I use universal language. It is because the dissenters from this generic Theism are so few as to be absolutely inappreciable in the presence of the empires and continents and generations who hold it with a profound and ineradicable faith.

What means this great Plebiscitum? What means this universal faith in at least one Worshipful Superhuman Intelligence — this chain of such faiths stretching away back into the mists of history and even the adyta of primeval tradition — this chain ever expanding toward Christian Theism as it passes through the more enlightened times and lands? If any man says that it means nothing, or that it does not flex itself significantly in the direction of God, my eyes dilate upon him with astonishment. Is he serious? Does he mean what he says?

It cannot be denied that universal and very ancient beliefs have sometimes proved false; but still it is acknowledged in practical life that they are generally true, and are always to be accepted as true in the absence of all positive evidence to the contrary. For example, if it should be the universal speech in this community that a certain person is dishonest, one would not, anterior to a thorough investigation, trust him as quickly as though there were no such common fame; especially if that common fame had existed for many years, and was fully indorsed by his most intimate acquaintances — proving that it is viewed as of the nature of evidence. It is possible that the man has been belied, for many instances of such belying have been proved; but still that universal faith against him is one of the adverse probabilities needing to be offsetted and overcome by other probabilities. In the absence of all discoverable positive evidence to the contrary, the universal and stable belief would be considered decisive against the man for all practical purposes, and ought to be so considered. Is there any positive evidence that there are no superhuman intelligences? On the contrary, are they not rather favored by the fact of numerous orders of living beings below us in a long line of gradation down to microscopic life? What authority has man for saying that the long line in its ascent ends with himself, or ends anywhere short of a Being of infinite proportions as compared with ourselves?

Further, there cannot be shown an instance of dateless and universal belief which has maintained its ground without abatement amid all advances of knowledge and morals, and which has even been enhanced by such advances, proving false. The false belief that the sun moves around the earth was universal at one time; but as knowledge increased this sort of astronomy weakened and passed away. The false belief in astrology, in the lunar influence on the weather, is very ancient, and has had almost universal acceptance; but it has faded before advancing intelligence. The false belief that it is lawful to worship many deities, and to represent deity under material forms, was for ages well-nigh universal; but wherever at any time knowledge and character have improved, polytheism and idolatry have shown tendency to decline. See, for proof, the French Positivists. But French Positivists were hardly needed to prove this to any moderate reader of history. The chief Greek and Roman philosophers seem to have always lived on or within the verge of Monotheism, spiritual Monotheism; and the more learned and better class of Brahmins at the present time, when drawn into explanations, take up very much the same position. — On the other hand, this dateless and universal belief in at least one Superhuman and Worshipful Intelligence has not been injured anywhere by a combined advance in knowledge and character; but the reverse. The Mohammedan nations, as such, believe as strongly as the pagan — the Christian nations, as such, as strongly as the Moslem — the most advanced Christian nations as strongly as the least advanced. So far, indeed, from this belief declining with advancing intelligence and virtue, it shows in such case a general tendency toward a more refined and stupendous Theism. Osiris, Jupiter, and Brahma, are far greater deities than any worshipped by African or South Sea savages — the Theos and Deus of such philosophers as Socrates, Plato, Cicero, Seneca, far greater than the popular Jupiter — Allah and Jehovah far greater than the divinity of Plato's speculations — even Jehovah as conceived by the cultured and saintly christian is a far more glorious object than the average Jehovah of Christian lands. In such lands those communities which are the most eminent for intelligence and excellent living are also most noted for both the strength and quality of their faith in the supernatural. See that swearing, swindling, drinking, gambling, dissolute, and ignorant frontier settlement! Which has the strongest and highest faith in the supernatural — that, or yonder cultured and virtuous New England village! See that good man of to-day! Make sure that when, twenty years hence, he has become a still better man — more solidly principled, more strictly conscientious, more loftily just, more tenderly and actively benevolent — his faith in God will stand on a still broader base and pierce the heavens with a still loftier apex. It is simple experience. Never stood pyramid more stably and sublimely than stood the faith of Sir David Brewster at the age of fourscore and seven — a faith that had grown through the long years as fast as his ever-growing intelligence and goodness.

Look at it. A dateless and universal belief in at least one Great Intelligence of a grade indefinitely superior to the human — whence came this mighty epidemic? Did it spring naturally from a low moral and intellectual condition of the race at large — as fevers and *ignes fatui* do from marshes — or from the selfish efforts of governments and incipient priesthoods; or from both? Either origin would be inconsistent with the fact that a combined advance in knowledge and morals is found to affect the faith favorably. Did it spring from the evident profitableness of the faith in the sight of all mankind? This were strongly in its favor as being true. Did it spring from the fact that it is intrinsically and universally palatable, if not profitable? Who can say that? No-Religion makes no exactions whatever: the easiest religion known to men makes great exactions, and makes them constantly. Self-restraint and sacrifice are the common and statute law of every religious system. Not a worship but includes endless expenses, labors, cares, and fears. Codes of regulations must be carefully studied out, and watchfully conformed to. Pilgrimages, penances, works from the twelve labors of Hercules downward, must be accepted. Temples must be built, altars fed, costly rites maintained, priesthoods supported. In fine, to mere nature, a religion is a cramping formula for this world; while it offers for another world only what, according to the atheistic theory, a man is equally at liberty to expect without a God. So it would seem to be intrinsically an unpopular system. That such a system could have fought its way from nothing into virtually universal acceptance, and maintained itself there unfalteringly from immemorial antiquity to the present, without any real support from either the reason, the experience, or the interest of mankind — could even have brightened and ascended with advancing knowledge and morals, and all as the product of the hideous incubation of wickedness upon general ignorance and wickedness — is, to say the least, far from being a plain matter. It has a strong look of incredibility. It savors mightily of self-contradiction. Plainly, it would take more argument than most minds can compass to give even plausibility to such an explanation. As to demonstrating its adequacy, such a thing is out of the question. The very idea is absurd. But if we suppose a primeval revelation of God; that the doctrine was gradually lowered and corrupted to a great extent by the moral and intellectual lapses of the race; that nevertheless it commended itself so mightily to their fundamental instincts, essential reason, and great wants that even such potent sources of error could never quite overpower it among any considerable body of mankind; and that, just as soon as these incubi are lifted, the elastic and irrepressible doctrine proceeds to expand toward its normal and original grandeur — I say, if we suppose this, we have an explanation of the general faith in worshipful superhuman beings, and of its obvious partiality for intelligence and virtue, which is perfectly natural and perfectly sufficient; intuitively so. The adequacy of the explanation is perfectly axiomatic. Not a word need be said in its defense. Especially in view of the fact that all the most eminent mythologists of the present day are agreed in the opinion that Monotheism lies at the foundation of all pagan mythology.

No one who in these times and lands admits wonderfully superhuman beings, but will go further, and admit a God. As a matter of fact, those who admit them do invariably admit a true God. And it ought to be so. For this admission takes away, on the one hand, the only serious appearance of an objection to a God, and, on the other hand, vastly intensifies the difficulty of accounting for Nature without Him; indeed, makes such an account impossible, if we may trust the mathematical doctrine of chances. The only apparent objection to a God that has much weight with most persons, is His failure to manifest Himself in overwhelming appeal to our senses and experience; and this objection is recognized as invalid just as soon as one admits any invisible intelligences above man who mingle in human affairs. And, too, just as soon as one admits such intelligences vastly above man and yet not eternal, he has introduced into the begun Nature that needs to be accounted for a new element of difficulty vastly greater than any it before contained. If it is somewhat hard to understand and show how blind causes can produce an intelligent man, it must be vastly harder to understand and show how such causes can produce an Intelligence vastly superior to man and able to make a man. In fact, the mathematics of chances forbids our attempting to account for Nature by blind causes after the admission of such a Being. La Place states the following law. The probability that an effect is produced by any one of given things is as the antecedent probability of that thing, multiplied by the probability that, if it existed, it would have produced the effect. Now, in the case before us, one agent is admitted as existing and able to produce the effect. To get the entire probability that it actually produced the effect, we must multiply certainty by the probability that, if existent, it would actually have produced the effect. Now the latter probability is certainly greater than the probability that a competing blind cause, if existent, would have produced it. It certainly is more likely that, of two causes, the one blind and the other intelligent, the intelligent was the author of an intelligent being or even of the human body. We know multitudes of organisms produced by intelligent beings, and not one certainly produced by blind causes.

Such is the Theistic hypothesis as compared with its sole rival. While perfectly sufficient, and, to say the least, *a priori* as credible as any, it is greatly the simplest; the surest; the sublimest; the safest; the most salutary; and the most in accordance with the convictions and traditions of mankind, especially of the most enlightened and moral part of mankind. In each of these respects it has almost infinitely the advantage over the law hypothesis. And, according to the maxims and practice of philosophy in other things, such an aggregate superiority as this ought to cause the Doctrine of a God to be promptly accepted and fully rested on as the true explanation of Nature. Whatever secular hypothesis could claim as much would be accepted without hesitation by all impartial men. It would be considered triumphantly established. To oppose it would be considered altogether absurd. And no man of science, with a reputation to lose, would for one moment think of venturing on opposition. On the contrary, an hypothesis so strongly fortified with verisimilitudes and superiorities over all competitors would ascend the throne of faith, and robe itself in the purple of all her prerogatives, by unanimous acclamation of the Baconian philosophy, of scientific usage, and of the entire college of scholarly men.

After the painting has been found pervaded with Titian's characteristics, you have only to observe that, as compared with other hypotheses in regard to its origin, that which attributes it to Titian is by far the simplest, the surest, the fairest, and altogether in accord with the convictions and traditions, especially of the best judges — I say, you have only to observe this in order to receive it cordially as the work of that old master. If able, you will give your thousands for it, on the strength of your convictions.

You believe that Canova made that statue, Angelo that cathedral, Herodotus that history. A neighbor has chosen to say that each of these wonders was made by a mollusk. This is his hypothesis. Another has chosen to say that each of these wonders was made by the great artist whose name it bears. This is his hypothesis. Why do you accept this last in preference to the other? Have you made out formal proof that the oyster cannot make such wonderful things — that though inert-looking things are sometimes found possessed of prodigious power, an oyster could by no possibility ever have wrought that shapely Venus, or swelled that surprising dome, or penned that immortal volume? Nothing of the kind. You do not deem such proof necessary. It is enough for you that the hypothesis which attributes St. Peter's to Michael Angelo is on its face altogether reasonable, that it has in its favor the whole current of tradition; while, as compared with the only competing hypothesis, it is almost infinitely the simplest, and surest as to adequacy. You have no occasion to inquire any further. It does not even occur to you to do it — cautious Baconian though you are. In common with the whole art-world, you instinctively accept and rely upon the great Florentine with unlimited boldness. The mollusk explanation is paraded before you in all sorts of ingenious verbal magnificence and logical forms without making the slightest impression on you. There is not a quaver in your faith. It not only occupies you, but reigns — not only reigns, but reigns indisputably.

So reigns to-day the Newtonian hypothesis of gravity. It is everywhere supreme — in the books, in the schools, in the innermost convictions of all intelligent men. Nothing moves wing, or opens mouth, or peeps against it. And yet do we see the principle of gravity? Not at all. Have we proved by experience that each particle of matter, away to the universe's last outskirt, attracts every other particle with a force proportioned directly to its own quantity of matter and inversely to the square of the distance between the particles? Not at all. Has it ever been demonstrated that the vortices of Descartes, or even the crystal machinery of Hipparchus and Ptolemy, cannot be so amended and appendixed as to explain all the astronomical motions thus far known? Not at all. Whence, then, that triumphant acceptance of the Newtonian principle and law of gravity? Simply from its superior merits as an hypothesis. Newton started a bare supposition. It was found to explain fact after fact. It kept on explaining. It has gone on up to the present time triumphantly explaining, in fields so broad, in fields so various, in fields so numerous and high, that our confidence in its power to explain the whole round of astronomical motions is quite complete. We deem it perfectly sufficient. Besides, while, to say the least, as credible on its face as the ancient Alexandrian or the modern French hypothesis, it is vastly simpler, surer, fairer, and more in harmony with the instinctive feeling and judgment of cultured men. This is the whole of it. This is the entire ground on which stand the entire scientific world. Is it not enough? Will any one start up at this late day to reprimand the entire scientific world for accepting and relying on the great Newtonian hypothesis of gravitation with unlimited boldness? And shall any venture to blame the theist for accepting and relying upon the Theistic hypothesis for precisely the same reasons somewhat intensified and enlarged? Confident as the astronomer may be that the clew which has not failed him yet in his wide terrene and stellar wanderings, would not fail him though his travels should go on to cover all the fields of Nature with foot-prints, still what he feels is not such a confidence as every sane man has that there is not a thing embraced by space whose origin the hypothesis of a God will not completely account for with infinite ease. This latter is the confidence of absolute, axiomatic, immeasurable knowledge. The other is merely the confidence of faith from a large induction of particulars. It is vastly probable — I consent to say morally certain — to at least philosophers, that this key of gravity will unlock the whole astronomical movement: it is mathematically certain to the entire sweep of humanity that this key of Theism will unlock and explain as to origin all the latitudes and longitudes of Nature. Further, in respect to simplicity, and sureness, and beauty, and accord with the convictions and traditions of mankind, especially the best part of mankind, the Theistic hypothesis has far more advantage over the law hypothesis than the Newtonian has over the Ptolemaic and Cartesian. And yet what a confidence the Newtonian displays! He threads great sciences on his doctrine like so many habitable globes. He sails away on his doctrine through the uttermost depths of heaven as on some voyaging sun. I will neither praise nor blame him. But this I say, that if he is warranted in founding himself so mightily on that doctrine of gravity, we are warranted in founding ourselves even more mightily on the Doctrine of God. We have the best of countenance in making Theism the basis of reasoning and action to any extent. We have no reproaches to fear from consistent science, though we proceed to rest upon that Theism castles, palaces, cities, empires, heavens of inferences and interests, answering to, but far nobler than, those which Astronomy confidently reposes on her great hypothesis.

A word more. For *what* would a man reject this vastly superior Theism? What does he gain by putting aside this account of Nature which carries itself so regally, and before whose sheaf all other sheaves bow down — the account which, while perfectly sufficient, and, to say the least, *a priori* as credible as any, is greatly the simplest, the surest, the sublimest, the safest, the most salutary, and the most suited to the convictions and traditions of mankind? Is he afraid of a personal God — lest that sharp-sighted Omnipotence should bring him to account for his conduct? Pray, in what respect would he be better off with a Nature constructed by law! Does the law scheme, necessarily, do away with sin? Does it do away, necessarily, with responsibility for sin? Does it, as a matter of course, even lessen the avalanche of penalty which the sinner may have to encounter? Not at all. All these things are just as possible, and may be just as great, under the one system of explanation as under the other. If a mist of atoms can really make this wonderful Nature which no man could make unless his faculties of wisdom and power were infinitely expanded — that is, if this mist seethes practically with an infinite efficiency, and its forces and laws taken together are fully equivalent, so far as the production of results is concerned, to that infinite power and wisdom which the Theistic hypothesis ascribes to God — then we have, to all intents and purposes, a material God. We have matter practically almighty and all-wise. It can do whatever an almighty and all-wise Person could do.

Now if men choose to call this wonderful thing by the name of Law, let them. If they choose to say it is unintelligent, let them. But let them not deceive themselves with names. What they actually have is something that can do things after a manner of unlimited wisdom and power. What they actually have is something that can arrange and adapt and exquisitely fashion just as if an infinite intelligence and discrimination, as well as force, presided over the work. In short, it is practically the equivalent of a God, if not God Himself. Such a Dynamic as this, whatever name it bears, is abundantly sufficient for everything. It can govern men as well as make them — it can treat them according to character as well as give them character — it can give us a glorious Bible in words as well as a glorious Bible in worlds — in short, it can do whatever Theism commonly attributes to God. Which is the harder — to make the arithmetical machine of Babbage, or to use it as it ought to be used? No, the Something that can make a man after a manner of infinite wisdom, can go on to deal with him, when made, after a manner of infinite wisdom. The potential Fog-Bank which is able to make men who can treat other men according to character, can itself treat them after the same manner of discrimination. So what do our atheists gain? What is their compensation for espousing the hypothesis that is the most intricate and far-fetched and uncertain and hazardous and hurtful and homely and hostile to the convictions and traditions of mankind? Their costly scheme — for the sake of which they are at the trouble and unreasonableness of such holocaust sacrifice of philosophy and taste and utility and venerable traditions — their costly scheme leaves men open to just as formidable possibilities as does Theism. The sinner has just as much reason to tremble before that astute Cauldron of mechanical and chemical forces that can make such a universe as this as he has to tremble before a personal God. Those are wonderful orbs yonder — this is a wonderful earth here with its packed life — even this single humanity of ours, body and soul, is an inexhaustible wonder to the most dynamical philosophy — full well do we know that the grandest man would have to develop into infinite proportions of intelligence and power before he could produce such an astounding universe as we behold — and, what I have to say is, that the primal Fire-Cloud which can organize such a universe as this which only an infinite man could organize, can, like such a man, practically discriminate between our righteousness and unrighteousness, and can, like him, pursue that unrighteousness as an unutterable Nemesis through all space and duration. Such a crafty Nebula is as fearful as God — only it can neither love nor be loved. It is as fearful as God to a sinner — though the atheist will never believe it, but will, while treating law as if almighty and all-wise for fashioning things, treat it as all-weak and all-foolish for the purpose of moral government.

AD FIDEM;

OR,

PARISH EVIDENCFS OF THE BIBLE.

BY THE AUTHOR OF

"ECCE CŒLUM" AND "PATER MUNDI."

ENLARGED EDITION.

Price $2.00 . . . . *Sent post-paid on receipt of price, by*

NOYES, HOLMES, & CO.,

117 WASHINGTON STREET, BOSTON.

"Ad Fidem" proposes to do for the Evidences of the Christian Religion what "Ecce Cœlum" aims to do for Astronomy. It proposes to bring these Evidences, without any sacrifice of scholarly accuracy, luminously and effectively within the reach of ordinary minds.

The attention of PASTORS is especially called to this work. Unbelief is trying hard to popularize itself. The most taking forms of literature are being used to insinuate doubt, and detach the masses from Church, and Sabbath, and Bible. Unless the shepherds of the people bestir themselves, a great calamity is at hand. They must see to it that what the friends of natural science are so finely doing for it, be done also for sacred science — that the Christian Evidences be brought to the people in those forms which alone are suited to interest and convince them. Cannot "Ad Fidem" help? If the judgment of men of the first eminence is worth anything, this is just the book needed for free circulation in the parishes.

EXTRACTS FROM NOTICES.

From Rev. Mark Hopkins, D. D., LL. D., President of Williams College.

This elegant volume seems to me admirably admirably adapted for its purpose. I am sure it cannot fail to do great good wherever it may go.

From Rev. Howard Crosby, D. D., Chancellor of the University of New York.

As a Christian minister, I thank the author of "Ad Fidem," *ab imo pectore*, not only for that book, but for all that he has done in his three noble works for the cause of truth. If the sympathy and approbation of his brethren all over the land is any reward for his labors, that reward he certainly has.

From Rev. Roswell D. Hitchcock, D. D., Professor in Union Theological Seminary.

Its bright, fresh, vigorous rhetoric, is one of the least of its merits. Evidently the author has himself felt, and so has justly measured, the "oppositions of science" which he combats. Only so can we get the confidence of thinking men, who are in trouble about the Bible. He does well to make so much of the moral temper of the inquirer. I often think that the apologetic literature of the Church, from first to last, has done little more than confirm and comfort those who were on the right side, and wished to remain there.

From Professor Taylor Lewis, LL. D., Professor in Union College.

I regard it as a very valuable contribution to our religious literature, and well worthy of the commendation the other works of the author have received. It is cheering to find that the many attacks on Christianity, under the names of science and free religion, are calling out so many books of intrinsic excellence. The great clamor of the enemy sometimes causes me to feel depressed; but such works as "Ad Fidem" assure me that there is power in the Church, both spiritual and intellectual.

From Rev. Austin Phelps, D. D., Professor in Andover Theological Seminary.

"Ad Fidem" has given me great satisfaction. It has been a greatly needed volume for a long while. What else have we in our literature on the Evidences which puts sound logic into readable style, so as to command the popular interest? I know of scarcely anything. Pastors are hard pressed, if I may judge from letters of inquiry which sometimes come to me, to find something which their inquiring young people will read by the side of the fascinating "Seers" of the Concord school. The author of "Ad Fidem" will find many to thank him for supplying the want.

From Hon. Jared B. Arbuthnot, LL. D.

Those who have known the author as one of the ablest mathematicians of the country; as a close student for years, and, almost to the sacrifice of life, of the profoundest branches of science; as a contributor to scientific journals of papers bristling with the utmost resources of the Calculus; and, latterly, as the author of a book on Astronomy, which has gone into many countries, drawn unprecedented eulogy from first scholars, and done more to make the most difficult of sciences intelligible and impressive to the general public than any other work ever written, will not expect to find him treating any subject superficially. They will not find him treating the Evidences in this manner. No reader of "Ad Fidem," who is himself a thorough scholar, will fail to see on every page of this, as well as of its companion volumes, under a popular dress, the order, thoroughness, immense force, and severe accuracy, as to both thought and expression, of a master in the exact sciences.

From Rev. A. P. Peabody, D. D., LL. D., Professor in Harvard University.

The author, or rather his numerous readers, should be congratulated on his continued and signal success in meeting the obtrusive skepticism of our times. His "Ad Fidem," in the choice and arrangement of topics, in its adaptation to existing needs, in soundness of reasoning, and in a vivacity and fervor which must command unwearied attention and interest, is precisely the work which the cause of truth demands. I am heartily thankful to him in behalf of the public for his service in the Gospel.

From Rev. W. S. Tyler, D. D., LL. D., Professor in Amherst College.

Clear as the air, bright as the sunshine, refreshing and invigorating as the northern breezes of this rare and beautiful season. There is in it a happy union of sound sense, good learning, personal experience, strong faith, and glowing eloquence, which bears the reader along as with an irresistible current. I admire particularly its boldness and directness. While there is sufficient moderation and prudence in stating the claims of the religion of the Bible, and the arguments by which it is supported, there is very little of the apologetic tone — there is no hesitation in appealing to the conscience and common sense of the *unbeliever* himself as on the side of the Christian Revelation.

I rejoice that the author has been permitted and enabled to add "Ad Fidem" to "Ecce Cœlum" and "Pater Mundi," and thus to lengthen and strengthen the chain which will, I trust, bind many to the truth.

From Rev. T. L. Cuyler, D. D., in the New York Evangelist.

Last evening my congregation enjoyed the intellectual treat of a brilliant discourse, by the author of "Ecce Cœlum" — that newly discovered star in our firmament of letters, in regard to whom so much interest is now felt. He is kinsman of President Burr, of Princeton College, and has devoted years to scientific studies. While listening to him, it seemed as if the frail form of flesh was ready to vanish away, while the inner soul was all aglow with the intense blaze of enthusiasm for the truth as it is in Jesus. His theme was — "The accord between the best literature and learning and the Word of God." It was a sparkling chapter from his newly published volume "Ad Fidem." The book abounds in sentences which are finished with the point of a diamond. Those who have read "Ecce Cœlum" will be hungry for this latest production of devout genius. The skeptic who can read its honest pages and not find his infidelity shaken, would hardly believe "though one rose from the dead."

From the Rt. Rev. Charles P. M'Ilvaine, D. D., D. C. L., Bishop of Ohio.

His admirable "Ecce Cœlum" had prepared the way in my house for its fit successor "Ad Fidem." In the range of its argument and in the force of its reasoning, added to the beauty and eloquence of its style, it is calculated to be, under the Lord's grace, eminently useful. The author appeals to evidences which none of the wisdom of the wise (of this world) can shake.

From the Springfield Republican.

"Ad Fidem" has met with much success — the first edition of fifteen hundred copies being exhausted within four days after publication. It is a vigorous and fascinating discussion of the Evidences of Christianity.

From the Interior.

The previous works of this author have been widely read, and much and justly admired. The volume before us is characterized by the same clearness and raciness, and will be read with interest by all classes.

From the Congregational Quarterly.

Dr. Burr has varied learning and remarkable rhetorical power. The earnestness and vigor of his faith are refreshing, particularly in an atmosphere surcharged with a speculative and skeptical spirit. "Ad Fidem" is well suited to relieve the doubts of the honest inquirer, and to strengthen the faith of the believer.

From the Literary World.

The author's fervor is exceedingly animating; the most indifferent reader cannot dwell unmoved upon his vigorous and glowing words; and those who reject his doctrines, must yield unqualified admiration to the skill and grace with which they are put forth. We have rarely fallen upon a professedly theological composition so rich in the genuine charms of rhetoric, so fascinating and persuasive in the delicate, yet forcible manipulation of grave and sombre subjects. Here is no dry discussion, no slow-going logical processes to disgust the reader with theme and thesis; the discussion is lively, the reasoning pleases while it convinces, and the impassioned earnestness of the writer allures his readers into willing tutelage, and brightens and beautifies his whole work.

"Ad Fidem" seems to us altogether admirable. It will bear and repay careful reading, for there has been no sacrifice of force to ornament. As a presentment of the claims of the Biblical religion, in a form at once universally intelligible and universally attractive, we know of no work which surpasses "Ad Fidem."

From the New York Observer.

"Ad Fidem" will, we believe, be greatly useful. It is admirably adapted to subserve the purpose designed. The author has made his mark as one of the ablest orthodox writers of the present day. He is a man of thought and study, and great power of expression. A short time since he burst on the religious mind of this country with a work called "Ecce Cœlum." He next appeared with a volume entitled "Pater Mundi," a profound, able, and timely series of chapters, proving that science testifies to the existence and attributes of the Christian's God. Modern professors of pure science would fain intimate to the world that it is unscientific to believe. Dr. Burr has made a book for these scientists and those who have been deluded by them to study. It is easy reading, and we recommend it to the learned and unlearned unlike. It will do them all good.

From the Christian at Work.

It is a worthy compeer of his two previous volumes. Rhetorically, it is most brilliant. It is full of passages which break upon the soul like a revelation, and in following the line of his arguments, the reader cannot fail to be convinced that of a truth the Bible *is* God's holy Word.

We welcome it as a most efficient helper in setting at naught the efforts which are being made to cast contempt upon the sacred writings.

From the Boston Journal.

Another valuable addition to the solid and beneficial literature of the day, from the pen of the well known author of "Ecce Cœlum," and the almost equally admirable "Pater Mundi." The present work is a most excellent one, calculated in every respect to accomplish great and lasting good. The

Evidences of the Christian Religion are brought within the scope of average intelligences. The book fills a most important place in the domain of modern religious literature. The style is graphic, powerful, and elegant; and yet beautifully simple. His arguments, though conclusive, are within the reach of the unlearned as well as the accomplished. Nothing hard or pedantic characterizes any one of the sixteen essays of which "Ad Fidem" is composed; but the book is pleasant and profitable reading for everybody.

From the Methodist.

Dr. Burr's previous volumes have rendered everything from his pen welcome to thoughtful readers. His new book consists of *real* parish lectures. It is a book of evidences skillfully wrought out, and the better for being popular. The author always presents a happy combination of scientific information with cogent logic and a vigorous style.

From the Religious Herald.

We welcome another volume from the vigorous and attractive pen of the author of "Ecce Cœlum." For weight of thought, brilliancy of imagination, and force of style, it will compare favorably with his former works; and this is enough to insure for it an extensive sale.

From the Home Journal.

This book will doubtless attract more general attention and be more widely read than any previous work from his pen. The writer's scientific habit of mind and familiarity with the whole field of argument have enabled him to give the proofs of revealed religion in a clear and forcible style, in a way to aid many who are seeking settled religious convictions.

From the Watchman and Reflector.

The author who, a year or two since, so greatly startled the reading public by vaulting into a first place among Christian apologists, is likely to hold what he so splendidly won. This last book is, like the others which preceded it, in the interest of the Christian Faith. The pages sparkle with life. Its poetic fervor, its wonderful massing of facts, its brilliancy of illustration, its personal appeals, its resistless conclusions, make up a book which will not allow the most prejudiced or indifferent reader to lay it aside, when once it is fairly begun, until the last page is turned. It is the most successful attempt which has yet been made at popularizing the Evidences of the Christian Faith.

From the Western World.

The work is spoken well and widely of as a strong defense of Christianity against the growing materialism of the age. Its author has a high reputation as one of the most powerful orthodox writers of the country.

From the Evangelist.

It presents the various branches of evidence in a very eloquent and effective manner. Moreover, it is peculiarly appropriate to the present state of the religious world — establishing the foundations of faith in the Word of God, and vindicating the supernatural character of the Gospel of Christ.

From the Scottish American Journal.

The author of "Ad Fidem" is already famous to the world by his admirable little book, "Ecce Cœlum." His books are probably more highly and universally extolled than those of any other author — not excepting the author of "Ecce Homo" himself. "Ad Fidem" will undoubtedly add to Dr. Burr's fame. It is a popular religious writing of the highest order, that can be read by the masses, and that will not fail to accomplish a good mission. This book of itself is calculated to turn the tide against infidelity in favor of the good old-fashioned belief in the Scripture as the Word of God.

From the Utica Observer.

Dr. Burr's "Ecce Cœlum" and "Pater Mundi" have placed him among the foremost of modern contributors to religious literature. As a Christian writer, his characteristics are great clearness, boldness, and enthusiasm. He seizes the sword of argument, and gives no quarter to limping skepticism that quibbles over the Bible as a book whose Divine origin is undemonstrable. His arguments are presented with remarkable vigor and they cannot fail to strengthen the faith of the weak, and to "confound the foolish," who accept as confirmed a thousand facts upon far less evidence than we have of the truth of the Bible as the very Word of God. It would be difficult to find a volume of three hundred and fifty pages of recent publication in which is combined more of sound logic and religious fervor, or which is likely to result in greater good than this. Dr. Burr is a man for the weak Christian to lean upon; for the strong and confident one to esteem and admire, if not indeed to reverence.

From the Commercial Advertiser.

This is a very welcome book from the pen of the distinguished author of "Ecce Cœlum" and "Pater Mundi." It is written at just the right time — at the time when the young men of the country show an unwillingness to "endure sound doctrine." Dr. Burr is a bold champion of the divine origin of revealed truth, and he handles skepticism without gloves. Let those who desire to know the truth read such a book as this. We do not fear the attacks of "scientists" upon revelation if those who read the speculations of science will, at the same time, exert themselves to reconcile history with Scripture prediction.

From the Advance.

A quite unanimous approval has greeted Dr. Burr in his labors as an author, as regards the value of his thoughts and the attraction of his style. The present work will meet with favor from those who appreciate the wants of our time. It aims to present the evidence in favor of the Bible, not in a dry, professional way, nor in a hot, polemic spirit, but with force and freshness, with appreciation of doubts and difficulties, and with the confidence of strong conviction. The author has much tact in coming at his subject, and his arguments are ingeniously constructed, and skillfully marshaled. He keeps in view, also, a practical result, and aims to impress the conscience as well as to enlighten the mind, insisting ever that the most solemn responsibility attaches to treatment of this great subject. We like the book, and wish it a large circulation.

From the Syracuse Journal.

Dr. Burr, the author, is a man in the prime of life, a lecturer in Amherst College, a man of profound scientific learning, patient study, and withal an earnest pastor, whose soul is aglow with enthusiam for the truth as it is in Jesus. His previous works, "Ecce Cœlum" and "Pater Mundi," have created a new sentiment in regard to religious subjects, and won for their author unbounded praise. They are notable books for the times, warm, alive, eloquent. "Ad Fidem" follows the path they marked out. In the words of Rev. Dr. Cuyler, "The skeptic who can read its honest pages and not find his infidelity shaken, would hardly believe 'though one rose from the dead.'"

From the North American Gazette.

The line of late publications indicated by "Ecce Cœlum," "Ecce Homo," etc., the first of which is from the same pen that now gives "Ad Fidem" to the world, can all be traced to the recent disputations in Europe over religious fundamentals. Of "Ecce Cœlum" we can hardly speak too highly to express the views of those concurring in its doctrine. It is thoroughly orthodox, compact, and thoughtful, and is a scientific as well as a religious essay; a work not unworthy to class with the great efforts of Chalmers. In half a dozen lectures it formulates more of the philosophy of orthodox faith than can be found in a century of ordinary sermonizing. This is the concurrent testimony of those whose opinions cannot be gainsayed.

"Ad Fidem" consists of a series of parish lectures, intended to settle the argument in behalf of the Bible. Of the execution of the labor too much can hardly be said. There is such an amount of plastic learning, close logic, and happy illustration, as justifies comparison with the astronomical discourses of Chalmers. Even the renown of Jonathan Edwards, so immovably crowned, is brought to mind by the closeness of the scientific analysis and synthesis used. And yet the whole is lucent to any ordinary understanding. The work takes instant rank with the foremost theological contributions of the day, and must exercise great influence.

From the Christian Recorder.

To secure the ready reading of "Ad Fidem" by those who have been fortunate to read "Pater Mundi," it is only necessary to inform them that it is from the pen of the same charming writer. It is a handsome book, and can be read with the most sensible joy.

It ought to be a question with thoughtful men, how these books of Dr. Burr can be placed in the hands of the people. We have not read "Ecce Cœlum," and consequently cannot speak personally of its worth. The others, however, we know to be books which the times demand. Could not cheap editions be issued — so cheap indeed, that the very widest circulation could be attained? With these in the hands of the class that make up the bone and sinew of the country, a strong bulwark would be erected against the rationalism of our German fellow-citizens, the papacy of our Irish, the infidelity of what few French we have, and the dizzy-headed nonsense of the few native-born Americans, who, to get notoriety, are willing to play the fool, in regard to the most vital of all subjects, religion.

From the Philadelphia Enquirer.

This volume consists of a series of lectures on the Evidences of the Biblical religion, delivered by Dr. Burr, the author of "Ecce Cœlum," a book which has gained a wide celebrity, in his parish in Connecticut. They were not originally intended for publication, but the author says that even if they had been they would hardly have been more careful in their statement of main facts and arguments. We do not think they would or indeed that they could have been much more exact or telling than they are. Dr. Burr is an advanced thinker, and a man of great liberality, so far as his books photograph him. His arguments are both cogent and persuasive, while through them breathes the all-powerful spirit of earnest conviction.

From the Congregationalist.

Some books are like a leaden rifle-ball; others like a cartridge, containing not only the ball but abundant means for propelling it. Dr. Burr's books are of the latter kind. This, his last, is not only a sound and good work, but it is active and stimulating. . . . We have a very able opening chapter entitled "Presumptions," which is worthy of being a book by itself so forcibly does it outline the grand general features of Christianity. . . . . Those who have read "Ecce Cœlum" and "Pater Mundi," will know what style to expect in the present volume. . . . . We accept this book as one of real power.

From the Lutheran Observer.

The readers of "Ecce Cœlum" and "Pater Mundi" — and their name is legion — will hail with delight this new work by the same "Connecticut pastor," who has so strikingly made the heavens declare the glory of God, and made the wondrous achievements of science testify to his wisdom, his greatness, his divinity and eternal power. It addresses itself to doubters and unbelievers with such an array of facts, and with such direct force of logic and argument, that it seems impossible for a rational soul to resist its conclusions. The book might appropriately be called rational and moral geometry, for its conclusions are the result of demonstrations as clear as any in Euclid.

The entire work characterized by great clearness and accuracy of style and statement, and it meets the objections and cavils of cultivated modern skepticism — the vague insinuations and sneers which float like froth upon the current of modern literature — better than other work that has yet appeared.

From the Christian Weekly.

"Ad Fidem" is a series of pastoral lectures to which the pastor has invited the reading public. And the reading public will be very apt to come when they learn that the lecturer is that same "Connecticut pastor" who fascinated them with the contagious imagination of "Ecce Cœlum" and "Pater Mundi." The same clear and cogent logic that in the former led us upon stepping-stones of stars to God as the father of the universe, the same glittering and brilliant style that in the latter led us through the phenomena of nature to God as the "Father of the World," is offered in "Ad Fidem" to lead us to God as our Saviour. With an air of confidence which betokens deep conviction; with an enthusiasm that is itself an evidence of Christianity, he insists upon the honest application to the Evidences of those tests which are prescribed by Christianity itself. And this is done with no juiceless language, but in a decidedly oratorical style, that will make the book very widely popular and useful. Its very fault — excess of ornamentation and gorgeousness of rhetoric — will secure a hearing for the truth by persons who might not be attracted by an ordinary book.

From the Evening Post.

We cordially thank the publishers for sending us this noble volume. It is most fittingly dedicated "To Christ and His Church." The work is full of irrefutable evidences of the Bible. In his delightful preface the learned and gifted author says, "Not only was Diderot," etc. The Typographical execution of the book is faultless.

From the New Englander.

Its merits are similar to those of his previous well known and popular volumes. The author has the gift of bold and impressive statement, ... a vivid and telling way of presentation, ... the glow and power of positive eloquence. The book will receive, as it deserves, extensive circulation, and, as we doubt not, will achieve great usefulness. We congratulate the modest and patient author upon the success which he has attained, and at which, perhaps, he himself is the most surprised.

From the Express.

The argument is strong and the style in which it is stated clear and impressive. The author is well known as one of the ablest and most interesting of the religious writers of the day.

From Harper's Monthly.

It is refreshing to come across a book written in a tone at once so candid and so cheeringly confident as "Ad Fidem." We find throughout the book, as Dr. Burr in his preface advises us we shall, "an air of great confidence." At the same time the author rarely substitutes mere assertion for argument, and never denounces as criminal the reader who fails to appreciate the force of his statements, and to accept the opinions to which they lead.

From the Princeton Review.

In this beautiful volume Dr. Burr expatiates in his favorite field of Apologetics with vigor, tact, and eloquence. He rightly traces the fortress of unbelief in the intellect to perverseness in the will and heart; shows that the difficulties in the way of religious belief are no more formidable than men encounter in every sphere of life without being stumbled by them; that with like candor applied to Christian truth all their embarrassments will vanish, etc.

From the Christian Quarterly.

These lectures discuss some of the living questions of the age in a manner at once able, pleasing, and practical. But we need not say this to those who have read *Ecce Cœlum* and *Pater Mundi.* These will know that it is almost impossible for Dr. Burr to write a dull book. *Ad Fidem* will *add* to the author's reputation. It is emphatically a book for the times; and is one of the finest defenses of the Christian religion that has been made in this country. It does for the Evidences of Christianity what *Ecce Cœlum* does for astronomy.

From the Baptist Quarterly.

This is a new work by an author who has achieved a popularity as widespread as it is merited. Dr. Burr writes in a style of singular freshness and vigor, groups his truths with great power, and communicates his enthusiastic earnestness to his reader.

From Scribner's Monthly.

The Rev. Dr. Burr, of Lyme, Conn., has made a sudden reputation of late by two attractive — perhaps we might even say brilliant — books on the Evidences of Christianity. He has just published a third. *Ad Fidem* is a rapid, popular, and eloquent summary of the arguments for the Bible. It is founded on careful research, and is believed to represent the latest developments of Biblical scholarship. There is no pretense of originality or appearance of scholastic learning; but the author has what is much better for his purpose, a forcible style, a dexterity in the use of striking figures and examples, and a remarkable gift of seizing and retaining the interest of his readers. He is clear, earnest, rapid, vigorous, and, above all, entertaining.

omy, as a whole, from it than I ever got before from all other sources — more than from Enfield's great book, which I once carefully worked out, eclipses and all.

"I trace the progress made, and the methods of the same, and seize on the exact status of things at the point now reached."

From the Bibliotheca Sacra.

"This is a remarkable book, — one of the most remarkable which has proceeded from the American press for a long time. It lifts the reader fairly into the heavens and unveils their glories. The presentation is very full though concentrated, very clear and animating, — with a command of language and a glow of eloquence which is quite extraordinary. The last lecture is hardly less than a Te Deum. The only adverse criticism which, on reading the preparatory lecture, we were inclined to make, was, that the almost impassioned eloquence with which it opened would have been more impressive further on, and after the imagination had been excited by the facts. But, after finishing the last Lecture, we could not wonder that a mind so full of the great facts, and of the emotion which they necessarily kindle, should, on seeing his own parish charge assembled to listen, break forth in strains which none but a mind fully roused by his theme and his audience would have been able to utter. No person can read through this volume without mental exaltation, and a conviction of the peculiar ability of the author."

From the New Englander.

"It presents an admirable *résumé* of the sublime teachings of Astronomy, as related to natural religion, — a series of brilliant pen-photographs of the Wonders of the Heavens, as part of God's glorious handiwork. The first five lectures pass the science in rapid review; the last treats of the Author of Nature, as related to its leading features. There is not a dry page in the volume, but much originality and vigor of style, and often the highest eloquence. It is, withal, evidently by an author at home in his subject, not "crammed" for the task. It affords a fine example of what an intelligent pastor can do, outside of his pulpit, towards training an intelligent people, and by imparting to them Nature's teachings, leading "through Nature up to Nature's God," — the God of Revelation as well. To such a book the author need not hesitate to affix his name."

From Rev. A. P. Peabody, D.D., LL.D., Preacher to Harvard University, and Plummer Professor of Christian Morals.

"Permit me to thank you for a work in which you have effected a rare union of scientific accuracy, eloquent diction, and rich devotional sentiment. It is attractive, instructive, and edifying. It appears at a time when science needs, as never before, to be redeemed and sanctified by faith in Him, in whom are hidden all the treasures of wisdom and knowledge. And, best of all, it does not make Religion cringe to Science, but maintains her in that queenly status which is the only position she can hold. The book must do great good, and I heartily congratulate you as its author."

From Rev. S. H. Hall, D.D.

"Ecce Cœlum is much more than a book-success. It will be honored as a most timely and admirable treatise to put into the hand of thoughtful young people, to 'turn off their minds from vanity,' and lead them to God."

From the New-York Evangelist.

"This unpretending, though elegant little volume, gives a most admirable popular summary of the results of Astronomical Science. The author has evidently mastered his subject, and he has presented it in a most striking manner, adapted to the comprehension of the common reader, and enriched with pertinent illustrations. The book is perhaps the most fascinating treatise on the science which has been published of late years, ranking indeed in many respects with that of the late lamented and eloquent Mitchell. One of its excellencies is that it does not hide God behind his own creation.'"

From the Religious Herald.

"A New Book, and one that is a book, worth its weight in gold or diamonds, for it is full of gold and precious gems, — diamonds of law and fact, — truths beaming with celestial light. I speak of 'Ecce Cœlum,' from the pen of Rev. ENOCH F. BURR, D.D., of Lyme, Conn., published by Nichols & Noyes, Boston, a duodecimo of 198 pages. Mr. Burr modestly signs himself 'A Connecticut Pastor,' but some college has rent the vail and written out his full name, and added to it a D.D. So much the better for Connecticut and for the world. Such light as the book contains ought not to be under a bushel.

"These six Parish Lectures are a masterly, vivid, easy, sublime presentation of the enchanting facts of Astronomy. They are adapted to all classes, — the learned and the unlearned. The astounding glories of the skies are tempered to our humble eyes.

"Let all read the book, old and young. Let it be found in every school, in every library, and in every home where wisdom is invoked. Read it, and you will exclaim, what glorious light it sheds from the throne of God upon the lonely pathway of man!"

From C. H. Balsbaugh, of Pennsylvania.

"It is certainly a wonderful little book. How the world shrinks into an atom as we follow the lofty soarings of the 'Connecticut Pastor.' I never knew rightly what Dr. Young means by saying, 'an undevout Astronomer is mad;' but I now see and feel the power and beauty of the expression. Such a book cannot be read without laying upon us the responsibility of a new charge from heaven. After contemplating such grandeur, we instinctively exclaim, 'What is man that Thou art mindful of him?'"

From Hon. S. L. Selden, Late Chief Justice of New York.

"A beautiful book. I admire it for the elegance of its style, as well as for the lucid and able manner in which it presents the noblest of the sciences. It will prove, I think, very valuable, not merely for the knowledge it communicates, but as suggestive of a line of noble and elevated thought. And I am much pleased to see from the numerous notices which have come under my observation that my estimate is confirmed by many persons of the first capacity for judging. To have written a work which receives and deserves such *very* high praise from scholars and men of science cannot but be a source of great gratification to the author."

ECCE CŒLUM;

OR,

PARISH ASTRONOMY.

ELEVENTH EDITION.

SUPPLEMENTARY EXTRACTS.

From the Theological Eclectic, [*Edited by Professor Day, Schaff, etc.*]

"The style is remarkably graphic and elastic, and the matter is so skilfully grouped and lucidly stated as to be level to all classes of readers. The writer has a rare gift at popularizing science, and his book deserves the wide welcome it has received."

From the New York Observer.

"We have never yet seen a volume on Astronomy that seemed to us to explain more intelligently, to ordinary minds, the visible phenomena of the heavenly bodies."

From the Congregationalist.

"We advise all our readers who have not yet read the book entitled 'Ecce Cœlum,' to embrace their earliest opportunity to do so,—a book which certainly has been surpassed by nothing of this general line, for many years, if ever. There is a grandeur of conception—an easy grasp of great facts—a clear apprehension of deep and subtle relations—a power to see, and make others see, the nature and extent of the heavenly movements, such as are altogether wonderful. Many works have been written from time to time to popularize astronomy—to bring its great leading features within the compass of unscientific minds. But we do not know of a work in which this has been so finely done as in 'Ecce Cœlum.' Six lectures of about an hour each, tell the story, and the reader feels, all the while, as if he were upon a triumphal march. He is upborne and sustained by his guide, so that he has no sense of labor and weariness on the journey. The last chapter, on 'The Author of Nature,' is a most worthy and fitting close to the book. We wish it could be read by that great host of so-called scientific men, who are delving away in the mines of nature, with thoughts and purposes materialistic and half atheistic. They need the tonic of such Christian thinking as this."

From Hours at Home.

"This little book, from the pen of Rev. E. F. Burr, D.D., has already been noticed extensively and pronounced a 'remarkable book' by our best critics. The author first delivered the substance of it to his own people in familiar lectures. It presents a clear and succinct resume of the sublime teachings of astronomy, especially as related to natural religion. The theme is an inspiring one, and the author is master of his subject, and handles it with rare tact, and succeeds as few men have ever done in giving an intelligent view of the wonders of astronomy, according to the latest researches and discoveries. It is indeed an eloquent and masterly production."

From Harper's Monthly.

"The title page of 'Ecce Cœlum' is the poorest page in the book. We have seen nothing since the days of Dr. Chalmer's Astronomical Discourses equal in their kind to these six simple lectures. By an imagination which is truly contagious the writer lifts us above the earth and causes us to wander for a time among the stars. The most abstruse truths he succeeds in translating into popular forms. Science is with him less a study than a poem, less a poem than a form of devotion. The writer who can convert the Calculus into a fairy story, as Dr. Burr has done, may fairly hope that no theme can thwart the solving power of his imagination. An enthusiast in science, he is also an earnest Christian at heart. He makes no attempt to reconcile science and religion, but writes as with a charming ignorance that any one had ever been so absurdly irrational as to imagine that they were ever at variance."

From the Evangelist.

"We have had many inquiries in regard to the authorship of Ecce Cœlum,' the volume noticed somewhat at length two weeks since. To save writing a number of letters, we may say here, that the Country Pastor, who is the author of these six Lectures on 'Parish Astronomy,' is the Rev. E. F. Burr, D.D., of Lyme, Ct. The book is a 16mo of about two hundred pages, but in that small compass it comprises the results of long study, and will be found as instructive as it is eloquent. The grandest truths are made level to the plainest understanding. We took it up, expecting little from its humble pretensions, but soon found that it was all compact with scientific knowledge, yet glowing with religious faith, and were not surprised that Dr Bushnell should say he 'had not been so fascinated by any book for a long time — *never* by a book on that subject' — and that it had given him 'a better idea of astronomy than he ever got before from all other sources.' We don't know if they have many such ministers 'lying around' in the country parishes of Connecticut, but if so it must be a remarkable State.

"While the impression of this fascinating volume is fresh in mind," etc.

From Rev. G. W. Andrews, D.D., President of Marietta College.

"The author has succeeded admirably in his attempt to present the great facts of Astronomical Science in such form as to be intelligible to those who have not gone through with a thorough mathematical training, and to make them intensely interesting to all classes of readers. I cannot express more strongly the interest the volume excited than by saying that I read through at once. I can hardly remember when I have done the same with another work."

From Rev. Edwin Hall, D.D., Professor in Auburn Theological Seminary

"I received it last night, and have read it through with intense interest and delight. It is a worthy book on a mighty theme. I wish it might be in every household, and read by everybody. And I am sure it will be read with admiration and wonder long after the author shall have been gathered to his fathers."

From Rev. Prof. E. W. Hooker, D. D.

"The book is an admirable argument from the discoveries of modern Astronomers, for the existence of God; and indirectly for the truth of the Gospel. It is an honor to his kindred, to the

Church and the place of his birth, and, above all, to Him whose gospel he preaches."

From an Obituary of Rev. S. L. Pomroy, D.D., late Secretary of the A. B. C. F. M.

"He was a man of extensive information, a ripe scholar, and he retained his scholarly habits and tastes to the last. A few weeks since he read 'Ecce Cœlum' with great pleasure and satisfaction, When he returned it he remarked, 'I have read it all twice, parts of it three times, and have noted down certain passages.' He was specially delighted with the arrangement of the work — the grouping of the different system so as to give us something like a comprehensive idea of the grand whole."

From the Congregational Quarterly.

That a Connecticut Pastor should be able in six lectures to his peoble to shed more light on this profound subject — to make it more simple and yet more grand, amazing, and impressive — than many of the great masters who have written before him is a matter of surprise. Yet this seems to be the generally conceded opinion of the press. We hear but one testimony concerning Ecce Cœlum. Any intelligent reader of it can understand what before has been only a mystery. It is worthy of the widest circulation.

From the Lawrence American.

There is not a dry page in these six lectures; but the glories of the skies are presented in a most enchanting manner, vivid, popular, grand, and glowing. Young and old should read it.

From The Christian Union.

We can commend this book in the heartiest manner. It is one of the noblest examples of the moral uses of astronomy that have appeared since CHALMER'S astronomical sermons. Besides their intrinsic merit, these lectures show what may be done by a quiet pastor of a village church for the instruction of his people. Every preacher has not the equipment required for a course of scientific lectures: but "where there is a will there is a way," and much more might be done than is done in broadening a pastor's literary education and in raising the literary tastes of his people.

EXTRACTS FROM NOTICES.

From the Rev. W. A. Stearns, D.D., L.L.D., President of Amherst College

I have heard them with the deepest interest. They are so clear, so logical, so rich in illustration, so unexceptionable and beautiful in style, and so conclusive in the argument attempted, that I have profoundly admired them. Those gentlemen who heard them when delivered here, would, I am sure, from the comments which they made upon them, agree with me entirely in the judgment I have expressed. May the Great Being whose existence these lectures so nobly defend from the attacks of the foolish, though calling themselves scientists and philosophers, spare the life of the author and enable him to complete the full course of thinking on which he has so triumphantly entered and advanced.

From Rev. Prof. C. S. Lyman, of Yale College.

All whom I have heard speak of these lectures have expressed for them the highest admiration. In thought and diction they are worthy of Chalmers.

From Prof. Julius H. Seelye, Professor of Mental and Moral Philosophy in Amherst College.

It is with great delight that I have received the new book. I like, especially, its whole attitude respecting the question discussed; that it is so full of faith and so uncompromising. Atheism is as unworthy the intellect, as it is repugnant to the heart; and I am tired of tame apologies from timid believers in a God. I like to see a book that has something of a clarion ring about it, and is not afraid to defy denial, when it speaks of the being and the glory of the Heavenly Father.

I believe that Pater Mundi will do great good, and I thank the Lord for permitting the author to prepare and publish it.

From Rev. A. P. Peabody, D.D. L.L.D., Preacher to Harvard University, and Plummer Professor of Christian Morals.

I thank the author with all my heart for Pater Mundi. It is the most efficient work of its class which the present generation has produced; and as the now existing scepticism is deeper, more [pseudo] scientific, more pretentious, than that of any preceding age; the book which, like Pater Mundi, is adapted to our times, must need be both broader and more profound than previous needs have elicited. Its treatment of the great theme is at once thoroughly philosophical and popular, both in style and in adaptation to the capacity of all readers of average intelligence. It was an unspeakable privilege to the students of Amherst College, to have heard the lectures; I trust that the same privilege will be extended through the press to thousands of our young men. While I find no fault nor deficiency in the treatment of any branch of the argument, I am especially impressed by the Seventh Lecture, as the clearest, strongest, and most eloquent statement of the need of God, and of the demonstration thence resulting of His existence, in the plenitude of His attributes, that has come within the range of my reading.

From Rev. Albert Barnes,

I was so profoundly impressed, or, if I may say so, *oppressed* and overwhelmed with the sublimity and grandeur of the truths presented in Ecce Cœlum, and with the manner in which the author presented these great truths, that I am glad he has followed with another volume on the same general subject. I anticipate in the perusal of it great pleasure and profit. I think the author is doing great service to the cause of truth and I hope that God will spare him to complete his work.

So far as I am able to judge, the greatest enemy which Christianity has to encounter now, is found in the oppositions of science, so-called. In fact, so far as I understand them, the aim and tendency of much of this science, are to blank Atheism; and I think a man can do no better service in this age, than to meet and counteract this tendency. I rejoice that God raises up men who are qualified to do it. I believe that the author of Ecce Cœlum is such a man. He has a noble work before him, and I hope he will be enabled to do it.

From the Independent.

We had not read *Ecce Cœlum,* and imagined that the enconiums which we had seen pronounced upon it must be too high wrought for sober truth. But now that we have read *Pater Mundi*, by the same author, we are ready to believe every word of praise to have been within bounds· The present volume is no dry, didactic treatise. It is warm, alive, eloquent. The author proves himself, in his freshness of thought and in the eloquence of his argument, inferior to no writer of the day. We find no slips in science, nor in his multiplied illustrations from ancient and modern literature. And we do find a grandeur of conception and a striking originality of conception, so audacious that scarcely any other writer we know of would have ventured upon it. We see no reason why our author's writings should not become classics in the language. Nothing can be more invigorating to the thoughtful reader.

From the Congregationalist.

We have read it with keen enjoyment, and are disposed to regard it as he most substantial and serviceable contribution to the natural theology of this generation, as it is the freshest and most popular. No better book none more entertaining, can be placed in the hands of inquisitive readers, especially bright minded young men and women· The author lays out his work with a singularly clear perception of the crepuscular skepticism which needs to be dissipated; and enters upon it with manly and generous fairness of statement, vigor of argument, and amplitude of apposite and convincing illustration. His style is in the main so admirable, that it may seem ungenerous to take exceptions. Probably the excess of ornamentation, the overfulness of illustration, the easy affluence of the most highly poetic diction, and the general gorgeousness of rhetoric will secure a hearing for the truth by persons whom it is desirable to influence, who might not be attracted by an ordinary book.

From the Hours at Home.

The decidedly oratorical style will serve to make the essays, incisive—eloquent, and eminently philosophical as we acknowledge them to be—all the more widely popular and useful.

From the Religious Herald.

Cogent argument is so lighted up with brilliant illustration, as to make interesting the profoundest thoughts.

From the Christian Union. Rev. H. W. Beecher.

The author, who, in *Ecce Cœlum*, established a reputation for that rare combination of excellencies—fervid rhetoric, scientific accuracy, and common sense—has produced another book designed to defend and illustrate the doctrine of Theism. It is like breathing mountain air to feel this man's earnestness; it is a true mental tonic. One sees instantly that he is able-souled, that he can push and climb without getting short of breath; and it is almost a foregone conclusion, after reading the first chapter, that one must either stride with him to *his* high conclusion, or part company before starting. This unequivocal earnestness and power display themselves at the outset; great heart is warmed up to begin with; so that one is almost inclined to distrust a leader who has so much the air of a partisan. The face set like a flint does not wait to be struck to emit its sparks, but glows with a fiery zeal which inflames everything it looks upon. Yet, no candid reader will say that DR. BURR is dogmatic; he only plies error with weapons for which infidelity has claimed a patent right. No one who reads this first volume, will wish that the author had written less or otherwise than he has.

From the Advance.

The previous work entitled *Ecce Cœlum*, received the highest commendation from the most competent judges. The present volume will still further augment the reputation of the author as a thinker and writer. He puts the Atheistic hypothesis to severe and annihilating tests; fully meeting its objections and cavils. The arguments of this work are not only cogent, but are expressed in a lucid, glowing, and eloquent style; and the book entitles the writer to a position among our best religious authors.

Fro ..ev. Edwin Hall, D.D., Professor in Auburn Theological Seminary

I have read the work with constantly increasing satisfaction and delight It is entirely worthy of the author of *Ecce Cœlum* and of its subject. So far as my reading extends—and I have long endeavored to read in that department whatever I could lay my hands on that promised to give me light—I regard it as the most original and valuable contribution to the subject, which the age has produced. I shall wait with longing for the second volume. In the meantime, I hope the work may have a circulation as extensive as its worth deserves. If it were left for me to fix that desert, there should not be a library or a family in the land without it.

From the Watchman and Reflector.

The thousands of readers of "Ecce Cœlum" have not got fairly over the feeling of astonishment and admiration which the perusal of that remarkable book brought to them, before another of equal merit from the same author is announced. "Pater Mundi," we are confident, will lessen nothing of the high character which Dr. Burr has won as an acute and accurate thinker, an accomplished scholar, a brilliant rhetorician, and a humble, childlike believer in God and His revelation. The purpose of the author is to defend and illustrate Theism and Christianity from the side of Modern Science. There is a wonderful candor in the entire process of argumentation. Nothing is assumed beyond what the eyes of man behold and his reason assents to. The conclusion, without being asserted, is irresistibly forced into one's own view, and wins acceptance from the thoughtful, reasonable soul. The eloquence of some of these passages respecting the fatherhood of God is overwhelming in effect. We earnestly commend the book to the careful study of our so-called scientific men who are trying hard to rule a personal God out of the universe. We wish, too, that every young man in the nation would read these pages. We are sure that nothing more fascinating in interest and really healthful and elevating in influence can be found among all the books of the day. The book is handsomely printed by Nichols & Noyes of this city.

From the Sunday School Times.

This volume is an eloquent and unanswerable protest against modern atheism in all its forms. "Modern science testifying to the Heavenly Father," is the author's secondary title, and it describes accurately the course and object of his argument. His methods of presenting the subject, however, are entirely original, and are wonderfully effective. The work is particularly opportune. There are in all our congregations thoughtful, cultivated, quiet men, whose faith has been shaken by the bold assumptions of infidel scientists. Dr. Burr's book is just suited to restore such persons to their equilibrium. It is written in a most attractive style and shows a masculine vigor of thought that cannot fail to command respect.

From the Theological Eclectic. Professors Day, Schaff, etc.

We have already spoken of the able work entitled Ecce Cœlum, in terms of high commendation. The present work by the same author exhibits the same power of comprehensive grouping and vivid presentation, and abounds in great thoughts freshly put.

From Rev. Mark Hopkins, D.D., L.L.D., President of Williams College.

I am greatly indebted to the author of Pater Mundi. It is a fresh and powerful work. If any commendation from me will aid its circulation, it is freely given.

From, C. H. Balsbaugh, Pa.

Certainly this is a book to stop the mouth of skeptics. It seems to me that never was atheism in its protean forms more squarely met on its own ground, and never more clearly discomfited with its own weapons. No two links of its argument are left together. The author has triumphantly vindicated the title of his book. Its matter and style appeal to both our innate susceptibility to truth, and our sense of the beautiful. In my view, never did logic and poetry more heartily embrace each other; never did beauty smile more divinely on the face of the sternest facts.

From the New York Evening Post.

The clear and beautiful logic, and the crystal style of Ecce Cœlum, fascinated religious minds everywhere in this country. This book is written by the same perspicuous pen. That it is in the form of lectures, rather improves it than otherwise. The special aim of the author is to wrest from the wild materials of this day the powerful sceptre of science, which they have seemed to wield. All the teachings of science and nature point to the "Father of the World." This book is one calculated to strengthen the faith of professors of religion, and to lead captive young minds straying into error. We ought to mention in closing, the beautiful typography of the book. Published by Nichols & Noyes.

From the Evening Wisconsin, Milwaukee.

The style is clear, and always strong and forcible in an unusual degree while many passages rise to great beauty and eloquence. Seldom have we read anything upon the subject of Christian evidence that was so entertaining, so instructive, and so satisfactory as this book. It is the offspring, of a vigorous intellect, and it is a most valuable addition to religious culture.

From the Christian Recorder, Philadelphia.

So charmed are we with this magnificent production of Dr. Burr's, that really we scarce know where to begin its praise. Its excellence is uniform

Lecture first and lecture eighth equally demand admiration. So every part of each lecture. The chain of gold is not only complete, but every link is complete. The Colonnade is not only symmetrical, but its minute carvings are perfect. To quote from it to our own satisfaction, would be to quote the whole book, but we remember that Messrs. Nichols & Noyes, the publishers, have a copyright.

How majestically does the author of Ecce Cœlum send forth his thoughts into the world! In majesty do they stride forth either to conquer, to convince, or to woo. Now as a mailed warrior are they seen, fully panopled from head to foot, and crushing by the strength of his arguments every foe—crushing every atheistic shield, and helmet, and breastplate. On almost every page of Pater Mundi, these all-crushing arguments are to be met—on almost every page we see victims lying mangled and bleeding.

We do not know that the author of Pater Mundi lays claim to the poetic gift; and yet has he given us a sublime Didactic Poem. Not in verse, is it given; it is neither Dactylic, Anapæstic, Iambic, nor Trochaic. But poetic imagination shines on every page. Untrammeled by rule, and enjoying a freedom that the utmost poetic license could not allow, the author has given us a poem infinitely sublimer than could possibly have been done in any other form. Would that we could give our readers the concluding pages of Lecture VII. Such poetic thought! Such beauty of expression! Such smoothness! Such harmony! Words answer to words, and sentence to sentence, with such sweetness that one glides along to the conclusion, as smoothly as a New England sleigh, and as merrily as its ringing bells.

From the Norwich Bulletin,

It will be a great advantage to the reader of this work to have made the acquaintance of Dr. Burr's previous volume, "Ecce Cœlum," as thus many of the references in "Pater Mundi" will be the more intelligible and vivid. The quality of the new work is in all respects admirable. Dr. Burr has a wonderful enthusiasm, always fresh and intense. He is full of his subject. He has the faculty of so treating profound and sublime themes, as to bring them easily to the comprehension of all. He has a fervid style, whose richness seems inexhaustible. He has great fertility in argument, and presents his suggestions with rare simplicity and force. The volume will go far to combat the sophistries of Atheism, both in uncultured minds and in those of strong logical powers. We cannot too highly commend it, and we predict that it will find a place in every well stocked religious library.

From the Standard, Chicago, Ill.

If any one should infer from the title of this book that it is a heavy and prosy dissertation, he would be astonished on looking over its pages

Nothing could be further from the truth. The author is an enthusiast, one of those who have not "discovered that one must be indifferent in order to be fair." The book is fresh, earnest, and eloquent, and we felt its strong spell before reading a dozen pages. The statement of arguments is admirably clear, the development of them is natural and impressive, and there is displayed a wonderful power in massing facts so as to give their full and combined effect.

From the Chicago Tribune.

This work in some respects is very remarkable. It is not only compact in argument, and forcible and clear in statement, but it is also absolutely brilliant and sparkling in manner, and rich and copious in illustration. Judging only from the one volume before us, we should pronounce it as one of the most remarkable and fascinating books of the day.

From the Orleans Republican, Albion, N. Y.

The author's premises are bold, and his line of argument clear, forcible and persuasive; shirking nothing, anticipating, and answering objections with equal fairness. The work is calm, liberal, and large thoughted; full of admirable logic, and profound reasoning; and the last three lectures, especially, are grand with beautiful and terrible imagery, exquisite poetry, and striking allusions to those mysterious facts and forces of nature which startle and awe believer and unbeliever alike; and his conclusion is singularly suggestive and powerful.

From Rev. Austin Phelps, D.D., Professor in Andover Theological Seminary.

I wish to thank the author for "Pater Mundi." Not that it needs any commendation from me: but I cannot but be grateful to any man who helps me to a new depth or vividness of conception of God; and this you have done by your book. I am specially impressed by the power with which it draws the great alternative, — a God benevolent, or a God malignant. The *reductio ad absurdum* is fearfully overwhelming; and the recoil with which one springs back from it gives one a lodgment and a resting-place in the Infinite Love which no gentler discipline could secure so well. This *vigor* of religious sensibility in your works charms me. We need it greatly in our Christian literature, to supplement alike the wiry intellect of which we have enough, and the emotive softness of which, perhaps, we have a little more.

From the American Baptist.

The author has a strong and vigorous style, and a power of grasping and grouping great truths, which make all that he utters luminous and convincing. Though prepared specially for educated men, they are adapted to all readers, have no abstruseness of diction, no intricate, far-fetched or dubious arguments. The author will impart no small measure of the indignation he feels towards atheism, concealing itself under the name of science, to those who read his book, and we trust it may have a very wide circulation.

From The New Englander.

The author of Ecce Cœlum could not well be expected to write a dull book on any subject, much less one in which God and nature were the chief topic. But whether he would be able to clothe a skeleton of a two-volume argument for Theism—often so dry and grim in other hands—with the flesh and muscle, the life and beauty, that charm us in Parish Astronomy, could only be shown conclusively by the production of a work like that before us. Pater Mundi, will, by the glow and magnetism of its rhetoric, and the enthusiastic earnestness of its tone, as well as the strength of its argument, be sure to command everywhere, appreciative and admiring readers, and prove, we trust, of special value to those who are inclined to regard science as hostile to religion. Its logic is vitalized and made effective by the force and richness of the illustrations drawn from the various fields of science. It is these all glowing often with poetic fervor, that rivet the attention at once, and carry the reader on insensibly from topic to topic. In some of the lectures, indeed, the argument assumes the elevation and almost the form of a grand poem. The sixth, for example, like a sublime ode, returns, strophe by strophe, with each point made in the argument, to the same exultant chorus, which becomes at once a ***quod erat demonstrandum*** to the understanding, and an inspiration of faith to the heart.

The second volume promises to be even more attractive than the first; for it is to be still more replete with the marvels and sublimities of the sciences as illustrative of the argument. It is too much forgotten by many that God may be studied in flower and forest, in storm and star, and in the soul of man, as well as in Moses and the prophets. The glowing pages of "Pater Mundi," teach impressively that the God of Revelation is the God of Nature as well.

From the Methodist.

The many gratified readers of "Ecce Cœlum." will welcome this new and important work of Dr. Burr. It is a book for the times. Natural Theology can no longer retain its old form: the progress, not only of Science but of speculative thought, demands a thorough revision, "Pater Mundi" meets this demand with masterly ability.

From the American Presbyterian Review.

A new work by the author of "Ecce Cœlum" is sure to attract unusual attention; nor will expectation be disappointed. Dr. Burr is an original and independent thinker, and he writes in a style of singular freshness and rhetorical beauty. His book is timely. Though popular in its address, it sacrifices nothing to effect, and is wholly free from that superficialty which is usually found in the attempt to reduce the conclusions of science to the level of a popular audiance. It discusses with masterly ability the testimonies of Modern Science to the being of a God, and defends Theism from the attacks of skeptical science in a bold and critical spirit worthy of all praise. It is as profoundly religious as it is thoroughly scientific. While it freely accepts the results of the freest investigations, it ably argues that there is nothing in one of these to shake the christian's faith, but much to confirm it. The work cannot fail to have an important influence on Natural Theology—bringing it into harmony with the progress of Science and speculative philosophy, and arming it with a new power of demonstration.

From the Princeton Review.

Dr. Burr, known to us in his youth as a modest and studious lad, and since, as the faithful and unpretending pastor of a rural congregation, has suddenly burst on our vision as an author of the first mark in the highest realms of thought, and as a leading defender of precious truth against the assaults of scientific pretenders and pretentious sciolists. He calls to mind the days when the great New England divines, the Edwardses, Bellamy, Backus, Smalley, Emmons, were pastors of agricultural congregations.

The universal approbation of Pater Mundi and the previous volume, by the press and by christian thinkers of the highest reputation, we find borne out by actual inspection. Real science is proved to be the handmaid of true religion, in a series of discussions which evince a masterly comprehension of the issues involved—a thorough acquaintance with modern science and its relations to religion—the whole in a style clear and simple, vivid and graphic. We think the quiet of a country charge more propitious to thorough study and deep thinking, than the din and whirl of metropolitan excitements.

From Prof. D. C. Gilman, Yale College.

I feel moved to express my hearty appreciation of the service the author of "Pater Mundi" is rendering to the world by the publication of these earnest, brilliant and impressive discourses.

From Hon. Henry L. Dawes, M. C.

The pleasure with which I read aloud to my family "Ecce Cœlum" has prepared me for an increased delight and profit in reading "Pater Mundi." I am very proud of the author, and rejoice in his growing fame.

From Our Monthly, Cincinnati, Ohio.

We are very glad to welcome and commend this book. The author does, with singular ability, what he proposes to do. His trumpet utters no uncertain sound. There is no danger of any one mistaking his meaning. We think it high time the arrogant assumptions and speculations of some scientific men in the interest of infidelity and atheism were exposed, and the harmony of all true science and revelation vindicated, made more apparent, and presented in some popular form. This Dr. Burr is doing, and the first installment of his work we have in this series of lectures. That they will be found interesting and convincing we need not say to those who have read ' Ecce Cœlum."

www.ingramcontent.com/pod-product-compliance
Lightning Source LLC
LaVergne TN
LVHW020222110826
845151LV00003B/796

* 9 7 8 1 4 2 5 5 3 2 7 8 9 *